# Innovation Addressing Climate Change Challenges

## CRITICAL ISSUES IN ENVIRONMENTAL TAXATION

**Series Editors:** Larry Kreiser, *Cleveland State University, USA*, Hope Ashiabor, *Macquarie University, Australia* and Janet E. Milne, *Vermont Law School, USA*

The *Critical Issues in Environmental Taxation* series provides insights and analysis on environmental taxation issues on an international basis and explores detailed theories for achieving environmental goals through fiscal policy. Each book in the series contains pioneering and thought-provoking contributions by the world's leading environmental tax scholars who respond to the diverse challenges posed by environmental sustainability.

Previous volumes in the series:
Original book published by CCH Incorporated
Volumes I–IV published by Richmond Law Publishers
Volumes V–VIII published by Oxford University Press
Volume IX onwards published by Edward Elgar Publishing

Recent titles in the series include:

Volume XIII Market Based Instruments
National Experiences in Environmental Sustainability
*Edited by Larry Kreiser, David Duff, Janet E. Milne and Hope Ashiabor*

Volume XIV Environmental Taxation and Green Fiscal Reform
Theory and Impact
*Edited by Larry Kreiser, Soocheol Lee, Kazuhiro Ueta, Janet E. Milne and Hope Ashiabor*

Volume XV Carbon Pricing
Design, Experiences and Issues
*Edited by Larry Kreiser, Mikael Skou Andersen, Birgitte Egelund Olsen, Stefan Speck, Janet E. Milne and Hope Ashiabor*

Volume XVI Environmental Pricing
Studies in Policy Choices and Interactions
*Edited by Larry Kreiser, Mikael Skou Andersen, Birgitte Egelund Olsen, Stefan Speck, Janet E. Milne and Hope Ashiabor*

Volume XVII Green Fiscal Reform for a Sustainable Future
Reform, Innovation and Renewable Energy
*Edited by Natalie P. Stoianoff, Larry Kreiser, Bill Butcher, Janet E. Milne and Hope Ashiabor*

Volume XVIII Market Instruments and the Protection of Natural Resources
*Edited by Natalie P. Stoianoff, Larry Kreiser, Bill Butcher, Janet E. Milne and Hope Ashiabor*

Volume XIX The Green Market Transition
Carbon Taxes, Energy Subsidies and Smart Instrument Mixes
*Edited by Stefan E. Weishaar, Larry Kreiser, Janet E. Milne, Hope Ashiabor and Michael Mehling*

Volume XX Innovation Addressing Climate Change Challenges
Market-based Perspectives
*Edited by Mona Hymel, Larry Kreiser, Janet E. Milne and Hope Ashiabor*

# Innovation Addressing Climate Change Challenges

## Market-based Perspectives

*Edited by*

Mona Hymel

*Arthur W. Andrews Professor of Law, University of Arizona College of Law, USA*

Larry Kreiser

*Professor Emeritus of Accounting, Cleveland State University, USA*

Janet E. Milne

*Professor of Law and Director of the Environmental Tax Policy Institute, Vermont Law School, USA*

Hope Ashiabor

*Associate Professor of Law, Faculty of Business and Economics, Macquarie University, Australia*

CRITICAL ISSUES IN ENVIRONMENTAL TAXATION
VOLUME XX

Cheltenham, UK • Northampton, MA, USA

Published by
Edward Elgar Publishing Limited
The Lypiatts
15 Lansdown Road
Cheltenham
Glos GL50 2JA
UK

Edward Elgar Publishing, Inc.
William Pratt House
9 Dewey Court
Northampton
Massachusetts 01060
USA

A catalogue record for this book
is available from the British Library

Library of Congress Control Number: 2018943994

This book is available electronically in the **Elgar**online
Law subject collection
DOI 10.4337/9781788973366

ISBN 978 1 78897 335 9 (cased)
ISBN 978 1 78897 336 6 (eBook)

Typeset by Servis Filmsetting Ltd, Stockport, Cheshire

Printed and bound in Great Britain by TJ International Ltd, Padstow, Cornwall

# Contents

# Figures

# Tables and boxes

## TABLES



## BOXES

# Editorial review board

The 15 chapters in this book have been brought to publication with the help of an editorial review board dedicated to peer review. The 18 members of the board are committed to the field of environmental taxation and are active participants in environmental taxation events around the world.

# Contributors

**Andersen, Mikael Skou**, Aarhus University, Denmark

**Ashiabor, Hope**, Macquarie University, Australia

**Astoria, Ross**, University of Wisconsin at Parkside, USA

**Aydos, Elena**, University of Newcastle, Australia

**Cantley-Smith, Rowena**, Monash University, Australia

**Chico-Almaden, Catherine Roween**, Xavier University-Ateneo de Cagayan, Philippines

**Grau Ruiz, María Amparo**, Complutense University of Madrid, Spain

**Gros-Désormaux, Jean-Raphaël**, The National Center for Scientific Research (CNRS), France

**Hymel, Mona**, University of Arizona, USA

**Johnston, Vanessa**, Monash University, Australia

**Kawakatsu, Takeshi**, Kyoto Prefectural University, Japan

**Kettner, Claudia**, Austrian Institute of Economic Research (WIFO), Austria

**Kletzan-Slamanig, Daniela**, Austrian Institute of Economic Research (WIFO), Austria

**Kreiser, Larry**, Cleveland State University, USA

**Lee, Paul J.**, Cleveland State University, USA

**Lee, Stephanie**, University of Scranton, USA

**Meier, Heidi Hylton**, Cleveland State University, USA

**Melendez-Obedencio, Marichu**, Xavier University-Ateneo de Cagayan, Philippines

**Milne, Janet E.**, Vermont Law School, USA

**Navarro, Ma. Kresna**, Xavier University-Ateneo de Cagayan, Philippines

**Nobrega, Bernardo Mendonça**, Lawyer, Brazil

**Palanca-Tan, Rosalina**, Ateneo de Manila University, Philippines

**Patel, Sejal**, Overseas Development Institute, United Kingdom

**Rubio-Sereñas, Caroline Laarni**, Xavier University-Ateneo de Cagayan, Philippines

**Rudolph, Sven**, Kyoto University, Japan

**Sprohge, Hans**, Wright State University, USA

**Tupiassu, Lise**, Federal University of Pará, Brazil

**van der Burgh, Laurie**, Friends of the Earth, the Netherlands

**Villar Ezcurra, Marta**, CEU San Pablo University, Spain

**Weishaar, Stefan E.**, University of Groningen, the Netherlands

**Whitley, Shelagh**, Overseas Development Institute, United Kingdom

**Worrall, Leah**, Overseas Development Institute, United Kingdom

# Foreword

A small handful of simple words might describe market-based instruments that address climate change policy: important, accelerating, interdisciplinary, and potential. Taken together, these words underscore the need to continue to study how market-based instruments do and can shape behavior in the face of climate change.

*Important*. While regulations play a strong role in reducing greenhouse gas emissions and adapting to the consequences of climate change, market-based instruments can send price signals that penetrate the complex networks of daily decision making, influencing behavior. Carbon taxes and cap-and-trade systems can harness the marketplace to achieve results. The unique characteristics of market-based instruments make them worthy of study and action.

*Accelerating*. Interest in using market-based instruments to address climate change has intensified rapidly in recent years. An electronic search of scholarly articles and books offers a rough proxy for the escalating trajectory of carbon taxes and cap-and-trade regimes in literature.[1] Works published in 1990 that mentioned 'carbon tax' numbered just over 100, rising to almost 800 published in 2000, then to over 4,500 published in 2017. Sources referring to 'cap-and-trade' rose from five published in 1990, to almost 200 published in 2000 and over 3,000 in 2017. On the heels of the growth in these terms, 'carbon pricing' came to serve as the conceptual umbrella for both carbon taxes and cap-and-trade schemes that cover greenhouse gas emissions. The number of works that mentioned carbon pricing in 2000 was only in the double digits, but over 2,000 works that were published in 2017 used the term.

The acceleration has not just been academic. Carbon pricing instruments have rapidly expanded in number and scope, as the World Bank documents in its recent report, *State and Trends of Carbon Pricing 2017*. In 1990, only two countries had implemented carbon taxes. By 2005, eight countries had adopted carbon taxes, and the European Union's Emissions Trading Scheme went into effect. Today, 67 national and subnational jurisdictions, constituting approximately half of the global economy, either have a carbon tax or cap-and-trade program in place or are implementing one. Market-based instruments are now mainstays in the portfolio of climate policies. With carbon pricing's heightened profile comes the need to constantly evaluate what is working and what is not.

*Interdisciplinary*. The growth in carbon pricing instruments to date is the result of legislators' determination to take on the challenges of addressing climate change and to correct market failures. It is also the product of efforts in many disciplines that have come together and built the foundation for political action. Science, of course, helps define the environmental goals. The design of instruments to achieve those goals must then draw upon economics. Economic modeling will shape the contours of pricing systems, and it will evaluate their success over time. Law contains the rules that govern the creation and implementation of new instruments. Public finance comes into play because market-based instruments can generate new revenue. Environmental policy helps evaluate the relative role of different policy instruments, including choices among different types of market-based instruments. Political economy informs decisions about what will be possible. Market-based instruments are the crystallization of many perspectives. Together, these perspectives create policies that can yield a better environmental result. Publications that assemble perspectives from multiple disciplines promote this synthesis.

*Potential*. Despite the trajectory to date, much remains to be done to realize the full potential for carbon pricing, whether in the form of carbon taxes or cap-and-trade regimes. The World Bank's 2017 report indicates that carbon pricing currently covers only 15 percent of global greenhouse gas emissions, a figure projected to rise to between 20 and 25 percent when China launches its national cap-and-trade program. And three-quarters of the covered emissions bore carbon prices that were less than US$10 per ton of carbon dioxide equivalent. Energy taxes also can put an implicit price on carbon dioxide emissions, as well as other environmental impacts associated with fossil fuel use. However, the OECD's new study, *Taxing Energy Use 2018*, finds that the effective tax rates on carbon from the combined effect of energy taxes and carbon taxes around the world are 'poorly aligned' with environmental costs. There is tremendous potential to expand the role of market-based instruments—and analyses of their role.

These four words yield a simple message for academic literature about climate policy. As carbon pricing moves from theory to practice, sharing analyses and ideas among disciplines and across national boundaries can enrich the understanding of experiences to date, the design of future policies, and the place of market-based instruments alongside other approaches. Few, if any, environmental problems are as fascinatingly crucial, complex and vexing as climate change. Market-based instruments will play a significant role. The challenge now is to ensure that they contribute as much as possible to the ultimate goal—reducing the effects and consequences of climate change.

Janet E. Milne
Professor of Law
Director, Environmental Tax Policy Institute
Vermont Law School, USA

## NOTE

1. The author relied on Google Scholar in performing this electronic search.

# Editors' preface

Although the world faces many environmental challenges, climate change continues to demand significant attention. Since the late 1800s, the average temperature of the earth's surface has increased by about 1.1 degrees Celsius. The 2000s have already seen record-breaking temperatures, with 16 years ranking among the 17 warmest global averages ever recorded. The change in climate has many consequences, of which rising sea level is just one. The rate of global sea level rise in the 2000s has been almost twice that of the previous century, when sea level increased by about 8 inches.

Recognizing this global threat, the 2015 Paris Agreement under the United Nations Framework Convention on Climate Change set the international goal of keeping temperature rise well below 2 degrees Celsius compared to pre-industrial levels. As a result of the Agreement, countries have committed to reducing global greenhouse gas emissions and have submitted nationally determined contributions that define their goals.

Market-based instruments can play a significant role in helping countries achieve their climate change goals. Volume XX of *Critical Issues in Environmental Taxation* explores a range of ways in which market-based instruments and complementary policies can play a role in helping countries meet their climate change goals. Its chapters demonstrate the potential that market-based instruments offer in reducing the risks of climate change, and also the challenges that exist. The chapters explore carbon pricing, as well as other tax and non-tax measures. Some examine existing environmental policies, others propose new policy ideas.

No single volume can fully cover the expanse of climate policy, given the complexity of addressing climate change and the multiplicity of existing and potential policies. We hope, however, that this volume will offer useful, market-based perspectives that can help inform the many climate policy decisions that lie ahead.

Mona Hymel, Lead Editor
Larry Kreiser, Co-Editor
Janet E. Milne, Co-Editor
Hope Ashiabor, Co-Editor

# Abbreviations

| | |
|---|---|
| ACCU | Australian Carbon Credit Units |
| ACU | Australian Carbon Units |
| AEMC | Australian Energy Market Commission |
| AGW | Anthropogenic global warming |
| BC | British Columbia |
| BDP | Bill Discount Pledge |
| BFAR | Bureau of Fisheries and Aquatic Resources |
| CCA | Community Choice Aggregation |
| CCS | Carbon capture and storage |
| CDD | Cooling degree days |
| CDFI | Community development financial institutions |
| CDG | Community Distributed Generation |
| CDO | Cagayan de Oro |
| CDORB | Cagayan de Oro River Basin |
| CDORBMC | Cagayan de Oro River Basin Management Council |
| CER | certified emission reduction |
| CET | Clean Energy Target |
| CF | Cohesion Fund |
| CFI | Carbon Farming Initiative |
| CMO | Carbon maintenance obligations |
| COWD | Cagayan de Oro Water District |
| CP | Carbon price |
| CPLC | Carbon Pricing Leadership Coalition |
| CPM | Carbon Pricing Mechanism |
| CPRS | Carbon Pollution Reduction Scheme |
| CSA | Community Service Administration |
| CVM | Contingent valuation method |
| DER | Distributed energy resources |
| DG | Directorate General |
| DOT | Department of Tourism (Philippines) |
| EAFRD | European Agricultural Fund for Rural Development |
| EbA | Ecosystem-based Adaptation |
| EC | European Commission |
| ECA | European Court of Auditors |

| | |
|---|---|
| EEA | European Environment Agency |
| E-GEN | Electricity generation |
| EIB | European Investment Bank |
| EITE | Emissions-intensive trade-exposed |
| EMFF | European Maritime and Fisheries Fund |
| END | Environmental Noise Directive |
| EPA | Environmental Protection Agency |
| ERDF | European Regional Development Fund |
| ERF | Emissions Reduction Fund |
| ESF | European Social Fund |
| ESIF | European Structural and Investment Funds |
| ETR | Environmental tax reform |
| ETS | Emissions trading system (EU) |
| EU | European Union |
| EURACOAL | European Association for Coal and Lignite |
| FEC | Final energy consumption |
| GATT | General Agreement on Tariffs and Trade |
| GDP | Gross domestic product |
| GGEs | greenhouse gas emissions |
| GHG | greenhouse gas |
| HDD | Heating degree days |
| ICAO | International Civil Aviation Organisation |
| ICMS | Imposto sobre a Circulação de Mercadorias e Serviços [Service and Merchandise Circulation Tax] (Brazil) |
| IEA | International Energy Agency |
| INDC | Intended Nationally Determined Contributions |
| IOU | Investor-owned utility |
| IPCC | Intergovernmental Panel on Climate Change |
| LIHEAP | Low-Income Home Energy Assistance Program |
| LMI | low- and moderate-income |
| LNG | Liquefied natural gas |
| LP | Liberal Party (Canada) |
| LPG | Liquid petroleum gas |
| LULUCF | Land-use, land-use change, and forestry |
| MBI | Market-based instrument |
| MFF | Multiannual Financial Framework |
| MILALITTRA | Miarayon Lapok Lirongan Tinaytayan Tribal Association |
| MoF | Ministry of Finance (Canada) |
| MS | Member States |
| Mtoe | Million tonnes of oil equivalent |
| NAS | National Academy of Sciences |

| | |
|---|---|
| NASA | National Aeronautics and Space Administration |
| NDC | Nationally determined contribution |
| NDP | New Democratic Party (Canada) |
| NEMA | National Emergency Management Agency (South Korea) |
| NGO | Non-governmental organisation |
| NOAA | National Oceanic and Atmospheric Administration |
| NPI | Non-productive investments |
| NRC | Nuclear Regulatory Commission |
| NYDPS | New York Department of Public Service |
| NYISO | New York Independent System Operator |
| OAR | Oro Association of Rafters |
| ODI | Overseas Development Institute |
| OECD | Organisation for Economic Co-operation and Development |
| OTDA | Office of Temporary and Disability Assistances |
| PES | Payment for environmental services |
| PPC | Public Power Corporation |
| PPP | Polluter pays principle |
| PV | Photovoltaic |
| RDP | Rural Development Programs |
| REV | Reforming the Energy Vision |
| RGGI | Regional Greenhouse Gas Initiative |
| RMU | Removal unit |
| SDG | Sustainable Development Goal |
| SM | Safeguard Mechanism |
| SME | Small and medium sized enterprise |
| SMI | Standard median income |
| TEV | Total economic value |
| TNSA | Tax on Air Transport Noise Pollution (France) |
| TPES | Total primary energy supply |
| UN | United Nations |
| UNFCCC | United Nations Framework Convention on Climate Change |
| VDER | Value of Distributed Energy Resources |
| VSL | Value of Statistical Life |
| WCI | Western Climate Initiative |
| WHO | World Health Organization |
| WTO | World Trade Organization |
| WTP | Willingness to pay |
| ZEC | Zero emission credits |

# PART I

# Carbon pricing design and prospects

# 1. Introducing carbon taxes – issues and barriers

**Stefan E. Weishaar**

## 1 INTRODUCTION

Carbon taxes in Europe are a relatively recent phenomenon. The introduction of carbon taxes can be subdivided in several phases as described by Andersen (2016).[1] They were first introduced in the 1990s in Finland, Sweden, Norway and Denmark, after the 1988 Toronto Conference on the Changing Atmosphere. These carbon taxes therefore coincided with a rising concern for global warming. The motivation for the introduction of this first wave of carbon taxes was, however, also related to the economic situations in these countries.

A second wave of carbon taxes was introduced in Eastern European countries such as Latvia, Slovenia, Estonia and Croatia. Taking place around the year 2000, the introduction of carbon taxes reflected the ambition to limit $CO_2$ emissions and to prepare for EU accession. Carbon taxes were a welcome source of additional income during difficult economic times.

A third wave of carbon taxes was enacted around 2010 in Ireland, Portugal and France, all countries experiencing budgetary challenges. The measures were motivated by climate change as well as fiscal ambitions though the revenues were modest in comparison to the countries' deficits. The participation of green parties in government (Ireland) or competition over environmentally minded voters (Portugal, France) eased the political acceptance.

The introduction of carbon taxes differed over time and country. Andersen[2] points out that it was the 'successful issue linkage' of non-environmental goals such as lowering payroll taxes, EU accession or revenue raising that provided the necessary political leverage for the adoption of carbon taxes. Environmental considerations were thus not the decisive factor. The challenges to be overcome have been country-specific.

This chapter examines the experience of the front-runner EU countries of the carbon tax (Denmark, Finland and Sweden) and addresses the question of what barriers to introducing the $CO_2$ taxes had to be overcome and how they were overcome. Knowledge on this is interesting as the Paris Agreement may lead to the proliferation of carbon taxes (e.g. in the Netherlands)[3] or to the strengthening of existing carbon tax systems.

The approach followed in this chapter is inherently multifaceted and takes economic and political aspects into account. It relies on a dual methodological approach employing a literature study with interviews. Interview partners were civil servants in the respective case-study counties who were selected on the basis of their experience. The chapter is structured as follows. Section 2 presents the economic framework prevailing in the 1990s and serves as a background on the introduction of the environmental taxes in Denmark, Finland and Sweden. Section 3 briefly presents the development of the $CO_2$ taxes. Subsequent sections address recycling (4), competitiveness concerns (5) and policy support (6). A conclusion will highlight the main findings.

## 2 ECONOMIC BACKGROUND

In the Nordic countries (Denmark, Finland, Norway, Iceland and Sweden) a special organization of economic and social policies emerged that is often described as a Nordic model or Nordic capitalism.[4] It is characterized by free markets with a comprehensive welfare state and collective bargaining at national level. It also features a combination of strong individualism and strong state, high levels of gender equality and social trust.[5] It has been argued that the strong individualism favours a strong support for market principles.[6] There is, however, not a single Nordic economic blueprint as the emphasis and approaches towards economic and social policies differed in the case study countries.

### 2.1 Denmark

The Danish economy is based on transport and agriculture. It joined the (now) European Union in 1973 and adopted a fixed exchange rate regime to address inflation though it has opted out of the Euro. As of 1982 the Danish Krona was pegged to the Deutsche Mark. This necessitated fiscal austerity measures, which in turn led to unemployment. As a consequence labour market rigidities had to be addressed. Expenditure for social transfers rose as welfare standards increased and unemployment soared during the mid-1970s to mid-1990s.[7] In the first half of the 1990s Denmark suffered from an economic crisis with high rates of inflation, large fiscal deficits and high unemployment rates. The low growth period was prolonged by the international recession in 1992. Danish unemployment figures peaked in 1993 at 10.1 per cent when the reform of the labour market compensation system showed effects.[8] In 1993 a new Social Democrat government decided to kick-start the economy by means of a moderate fiscal expansion while in 1994 the same government tightened labour market policies. As a consequence Denmark entered into a period of moderate growth with unemployment steadily falling.

## 2.2 Finland

The 1980s were years of high inflation and currency devaluations. The country had problems in controlling the credit market regulation and suffered from the collapse of the Soviet Union (an important trading partner) in the early 1990s. Unemployment soared and the currency policies had to be changed, leading to tight fiscal policies. Government debt in terms of GDP trebled in the few short years from 1990 to 1993. Under the impression of the political trauma created by the deep recession comprehensive reform programmes were implemented in Finland, leading to new macroeconomic policies, independent central banks, strict budgetary rules, deregulation and reductions in the welfare state.[9]

## 2.3 Sweden

The Swedish economy grew very slowly in the 1970s and 1980s and faced great structural challenges as its competitiveness declined.[10] Taxes rose and the welfare state expanded. The 1980s saw high inflation, currency devaluations and interest rates rose due to fixed exchange rate targets. These in turn led to a banking crisis, a severe economic recession, high unemployment and very high budget deficits.[11] Responses to recession (subsidization or nationalization) were ineffective and the focus shifted towards opening markets and embracing competition.[12]

By the early 1990s Sweden saw a severe economic crisis and the resulting political trauma facilitated comprehensive reforms to open up the economy and reduce the regulatory burden. Reforms extended to the tax system, new macroeconomic policies, independent central banks, strict budgetary rules, deregulation and reductions in the welfare state and the pension scheme.[13] The public debt burden doubled to 80 per cent of GDP during 1990 to 1995.

It can therefore be observed that by the 1990s the successful Nordic economic model came under stress and needed to deal with rising unemployment, competitiveness issues and increasing public debt.

# 3 $CO_2$ TAXATION

$CO_2$ taxes were not implemented in isolation but were often part of a larger environmental tax reform (ETR).

## 3.1 Denmark

Denmark introduced a carbon tax in the early 1990s. The $CO_2$ tax was not intended to increase the overall price on energy but to incentivize consumption of less $CO_2$-intensive energy sources and reflected increased climate change awareness.[14] It was introduced in multiple phases. In May 1992 it was introduced on energy products consumed by households. Industry that paid the $CO_2$ tax was refunded during 1992, thereafter businesses had to pay as well.[15] The tax was around 13 Euros per ton of $CO_2$.[16] From 1993 to 1995 industries had to pay only 50 per cent of the total $CO_2$ tax rate.[17] Based on the energy intensity of industries, even more generous treatments could be applied, reducing the overall tax burden to 10 per cent of the $CO_2$ tax rate.[18] This system was in force until 1995.

In 1993 the socialist Danish government proposed changes to the carbon tax provisions applicable to the business sector in order to ensure that the country would meet its climate policy target of reducing $CO_2$ emissions by 2005 by 20 per cent compared to 1988.[19] These changes for the business sector entered into force in 1996. The $CO_2$ tax applicable was now based on different type of uses. $CO_2$ taxes were highest for 'Industry space heating' and the 'Household and service sector', lower for 'Industry light processes' and lowest for 'Industry heavy processes'.[20] Companies could further reduce their tax burden if they signed an energy efficiency agreement with the Danish Energy Agency and invested in energy-saving equipment.[21] In 2005 the $CO_2$ tax was reduced to 12 Euros per ton of $CO_2$.

## 3.2 Finland

The first country to introduce a $CO_2$ tax was Finland in 1990.[22] The tax was levied on all energy products (light fuel oil, heavy fuel oil, coal, natural gas and peat) except transport fuels as these were already subject to energy taxes.[23] It was implemented as an excise duty on energy products. Over time the tax changed. Between 1990 and1994 the $CO_2$ tax was based on the carbon content of the energy product and set at around 1.2 Euros per ton of $CO_2$. Subsequently (1994–96) it was based on the carbon as well as the energy content of the energy product. The weighting started off as 60 per cent carbon content and 40 per cent energy content but subsequently changed to 75 per cent and 25 per cent respectively. In 1997 the tax changed back towards a full $CO_2$ tax. Over time it rose to 18 Euros in 2003, and to 20 Euros in 2008.[24] Thus the tax changed frequently and on an ad hoc basis despite the declared intention to have introduced a permanent $CO_2$ tax system.[25]

### 3.3 Sweden

Sweden introduced taxes on gasoline as early as 1924. Taxes on diesel, mineral oils and coal and liquid petroleum gas (LPG) and gas followed. In 1991 it introduced a tax on carbon emissions. It was part of a fiscal reform process primarily aimed at reducing labour taxes by increasing environmental taxes. The income tax reduction (4.6 per cent of GDP) was partially offset by the $CO_2$ and $SO_2$ taxes (1.2 per cent of GDP).[26] Energy taxes were lowered to compensate for the $CO_2$ tax. The $CO_2$ tax is based on the fuel content of the fossil fuel. In 1991 the tax rate was around 43 Euros and increased to approximately 100 per ton in 2007 and to 106 Euros in 2008.[27]

Subsequent tax changes were at times motivated by competitiveness concerns. As special tax reductions have not been granted to the Swedish industry, this led to an increase in the overall tax burden. Until 1993, industry and households had been charged the same high energy and $CO_2$ tax rates but the energy and $CO_2$ tax burden was dramatically lowered for industry, agriculture, forestry and fisheries in 1993 in the wake of the economic crisis. From 1993 onwards, these economic sectors were exempt from the energy tax payments and were only subject to a reduced $CO_2$ tax. Since 1998 the $CO_2$ tax rate for industry has remained constant in real terms.

As presented above, the $CO_2$ tax was first introduced in Finland. While the Finnish tax scheme was designed to be revenue raising it only placed a modest cost upon emissions. The $CO_2$ taxes in Denmark and Sweden were higher but unlike their Finnish counterpart quickly included substantial derogation schemes for industry. It is also noticeable that all $CO_2$ tax schemes were adapted on several occasions.

## 4 RECYCLING

This section of the chapter presents introduction issues and barriers from a comparative perspective. Recycling can be linked to ETR.

### 4.1 Denmark

Denmark reduced income taxes and increased environmental taxes, initially targeting households as industry was not affected by the 1993 tax reform. The second phase (1996–2000) of the ETR was smaller in magnitude and was more directed towards industry: employers' contributions to the labour market pension fund and employers' contributions to the Act on labour market funds were reduced and energy efficiency subsidy programmes and a special fund for small and medium sized enterprises (SMEs) were set up. The refund scheme was overhauled so that industry would bear the same energy taxes as households. An important feature of this phase is that there is no cross-subsidization between industry and households. In the third phase (1999–2002) environmental taxes and corporate taxes were increased in order to reduce personal taxes and taxes on the yield of pension savings and share yields. The tax burden in this phase fell most heavily on households as the reform especially increased the energy tax rate whereas the business sector was largely unaffected.

### 4.2 Finland

The Finnish $CO_2$ tax was motivated by both fiscal and environmental considerations and implemented as an additional excise duty on energy products. Recycling measures in Finland were only introduced several years after the introduction of the $CO_2$ tax when in 1998 energy-intensive firms could benefit from a tax refund system.

Prior to the introduction of the $CO_2$ tax, a political agreement was reached that the $CO_2$ tax would be introduced if income taxes were reduced in return. Also in the 1996 budget negotiations of the coalition parties, reform of the energy and $CO_2$ tax system was reached by agreeing on further income tax changes. The shift from environmental to labour taxes in Finland thereby pre-dates the actual recognition of an ETR as a policy tool. Even though Finland was the first country to introduce a $CO_2$ tax, it was one of the later countries to embrace an ETR.

### 4.3 Sweden

The 1991 tax reform in Sweden was directed to substantially reducing personal income taxes and was partially offset by changes in value added tax and the ETR. The reform was not intended to be revenue neutral. During the years 2001 to 2010 the Swedish tax reform emphasized the lowering of taxes paid by low and medium wage earners and the reduction of social security contributions.

Both Sweden and Denmark were following similar strategies and the $CO_2$ tax was part of a wider ETR reform. They both recycled money to finance income tax reductions and reductions in the social security payments paid by employers. Neither of them aimed at budget neutrality though Denmark emphasized that there should be no cross-subsidization between households and industry. In Finland the $CO_2$ tax introduction led to income tax reductions but not to an ETR. Finland did not recycle money to industry until many years after its introduction.

# 5 COMPETITIVENESS

Given the economic situation, competitiveness concerns were high on the policy makers' agenda in Denmark, Sweden and Finland.

## 5.1 Denmark

Competitiveness concerns were important in Denmark. When introducing the $CO_2$ tax, the business sector did not pay energy taxes and the 1992 $CO_2$ tax was fully borne by households.[28] During the years 1993 to 1995 non-energy-intensive companies enjoyed a lower $CO_2$ tax rate (50 DKK instead of 100 DKK) as well as a generous refund scheme that was dependent on the size of the $CO_2$ tax in relation to net sales. Under this scheme energy-intensive companies were refunded 50 per cent of the $CO_2$ tax paid in excess of 1 per cent of the net sales if the total amount of the $CO_2$ tax due was equivalent to 1 and 2 per cent of net sales. If the $CO_2$ tax was between 2 and 3 per cent of net sales the refund was 75 per cent of the amount exceeding 2 per cent. While if the $CO_2$ tax was above 3 per cent of net sales the refund was 90 per cent of the amount exceeding 3 per cent. This refund scheme reduced the average tax burden to 35 per cent of the standard household tax rate[29] and in effect placed a lower $CO_2$ tax burden on the manufacturing sector. Moreover additional tax support was available for up to three years if the company paid at least 10,000 DKK in $CO_2$ taxes.[30]

When the first phase of the ETR was introduced (around 1994) the Danish government had already announced that new environmental taxes would also be introduced for industry. For this an inter-ministerial committee was established which recommended that the $CO_2$ tax should be paid by industry, that tax rates should differentiate according to energy intensity and that revenues should be recycled back to industry.[31] These recycling measures took the form of lowering employers' social security contributions (reductions of employers' contributions to workers' pension funds and employers' contributions according to the Act on labour market funds) and investment grants for energy-saving measures. Moreover a fund for support of SMEs was created.

The second phase of the ETR mainly targeted industry and entered into force in 1996. Safeguarding the adverse effects on Danish competitiveness, all money that was levied from businesses would be recycled back to them.

The $CO_2$ taxes applicable to industry changed in 1995 and obliged companies to pay the $CO_2$ tax in accordance with usage. Regarding space heating, companies paid the same as households while regarding light processes, they paid 50 DKK, and as of 2000 90 DKK. The heavy processes tax rate was 5 DKK and as of 2000 25 DKK.[32] Heavy processes were those processes where the tax burden of 50 DKK per ton of $CO_2$ permanently exceeded 3 per cent of the value added of the enterprise, while the tax burden at the same time exceeded 1 per cent of the sales.[33] Very substantial $CO_2$ tax reductions were available for companies that reached an agreement with the Danish government on investing in energy efficiency.[34]

## 5.2 Finland

The Finnish $CO_2$ tax scheme did not have user-specific exemptions during the period 1990–96 and it is therefore not surprising that the nominal $CO_2$ tax rates in Finland are low by international standards.[35] During the period 1994–96, when the $CO_2$ tax was combined with the energy tax, there was a lower tax rate on natural gas and no $CO_2$ component was levied on peat motivated by energy and regional policy considerations.[36] Competitive considerations in the power sector and manufacturing were high on the policy agenda.

The change back to a 100 per cent $CO_2$ tax (1997) was motivated by criticism of electricity producers and large energy users.[37] The business environment of energy-intensive industries and electricity producers changed as Finland prepared for EU accession, electricity market reform and rising energy taxes – as a consequence industry was concerned about international competitiveness and questioned the environmental effectiveness of the energy and carbon tax regime in a common Nordic energy market.[38] Even though Finland had the lowest energy taxes in the Nordic countries, the tax on fuels for heat production was changed to a full $CO_2$ tax, the electricity tax was changed from a production to a consumption orientation and a tax refund system was implemented in 1998 for energy-intensive industries. Under this refund scheme 85 per cent of the amount paid in $CO_2$ tax and electricity taxes exceeding 50,000 Euros can be refunded provided that the total tax burden exceeds 3.7 per cent of the production value added.[39] Meanwhile the $CO_2$ tax was increased to compensate for the reduced fiscal income.

## 5.3 Sweden

The carbon and energy tax are closely linked and have to be assessed jointly[40] when addressing competitiveness issues. Initially the Swedish $CO_2$ tax did not provide for derogations for industry but increasing tax burdens led to competitiveness concerns as industry and households paid the same energy and $CO_2$ tax rates, though the total energy tax burden of companies was capped at 1.7 per cent of the sales value until the end of 1991 and as of 1992 it was 1.2 per cent.[41] This meant that year on year individual firms had to apply to the tax authorities, which was impractical, expensive and subject to criticism (nepotism and corruption) and potentially subject to challenges under e.g. World Trade Organization (WTO) law.[42]

The Swedish manufacturing industry was exempt from paying energy taxes as of 1993 and subject to reduced $CO_2$ tax rates. In the years 1993–97 it was 25 per cent, 50 per cent during 1998–2000 and subsequently reduced towards 21 per cent in 2005 (in 2001: 35 per cent, 2002: 30 per cent, 2003: 25 per cent, 2004: 21 per cent, 2005: 21 per cent). This helped to limit the overall tax burden (comprising energy and $CO_2$ taxes) for the Swedish manufacturing sector.

Energy-intensive companies benefit from a refund on their $CO_2$ tax if the tax due exceeds 0.8 per cent of the value of sales. In this case companies pay a reduced percentage amount over the excess tax burden. Energy-intensive companies whose carbon tax bill exceeds 1.2 per cent of the sales value are exempt from paying any tax on the excess amount.[43]

The impact on the business sector has to be seen in the context of the ETR, which encompassed taxes on tap water, waste water, plastic and paper bags.

In comparison to the comprehensive Danish tax exemption scheme on energy, the Swedish scheme seems more transparent and less elaborate in giving industry fewer possibilities to avoid excise duties on energy consumption. Implementation also appears to be simpler and cheaper as there are fewer exception options in the Swedish legislation. Moreover the Swedish scheme for energy excise duties is predominantly based on the $CO_2$ tax whereas the Danish scheme is oriented towards the energy tax. The tax design in Sweden places a higher tax burden on fossil fuel energy consumption by Swedish industry compared with Danish industry. Finland, by contrast, has long avoided support for industry and addressed competitiveness concerns by having a comparably low $CO_2$ tax level. Only as of 1998 were derogations for energy-intensive industries introduced, though particular energy sources such as peat and natural gas were enjoying a lower tax burden.

# 6 POLICY SUPPORT

## 6.1 Denmark

The implementation of the $CO_2$ tax in Denmark was made possible by a balancing of diverse interests of different groups of society. At the time of introducing the tax the centre party was in favour of taking climate change measures while this was not a policy priority for the two conservative parties. The political support for introducing the $CO_2$ tax was granted by the Social Democrats by earmarking parts of the tax proceeds for improvements in the Danish district heating system. The improvement and expansion of district heating was also a positive element in gaining support of the unions as this promised employment opportunities not only for workers, but also for union officials.

Another explanation for the political support for the $CO_2$ tax is that a $CO_2$ tax favoured the investments that had been made in the natural gas market. Due to the oil crisis prices for oil had been very high in Denmark and the political decision was taken to reduce the nation's dependency on oil. As oil prices declined, however, by the mid-1980s the (industrial) policy case for the decision to invest in gas appeared to be less compelling. Policy makers as well as utilities could find themselves supporting gas.

When introducing the $CO_2$ tax, competitive concerns (national and international) were an important point of consideration. Companies appeared at times to publicly welcome the $CO_2$ tax while requesting the trade associations to take a strong position against it. Public sentiment was becoming more environmentally minded and enterprises liked to be associated with responsible environmental conduct. A way to overcome this obstacle was by setting up a generous refund scheme that allowed companies to pay a reduced amount of $CO_2$ tax rates. The system was in effect placing a lower $CO_2$ tax burden on the manufacturing sector.

The support scheme for energy-intensive (heavy processes) companies was politically challenging to design given the high degree of technology diversity in Danish industry. The solution to focus on the value added of the enterprise constituted a limited administrative burden for companies as this information was already relied upon in other fiscal contexts. The industry support schemes at times offered companies the opportunity via restructuring of their organizational form to lower the $CO_2$ tax burden. An important element to gain support and make the $CO_2$ tax a success may also be found in creating the possibility to have agreements between companies and the government regarding energy efficiency improvements. Such agreements placed energy efficiency on the agenda of management and required management to pay more attention to their engineers.

Households were shouldering the major burden of the $CO_2$ tax. The introduction of the $CO_2$ tax was partly offset via a reduction in the existing energy taxes, which did not affect industry.[44] Positions of households may have been co-determined by the wider context of the Danish ETR, which was intended[45] to be a revenue-neutral tax shift programme and led to reductions in income taxes.

## 6.2 Finland

Finland is a sparsely populated country with long transport distances and an energy-intensive industry base.[46] It therefore took special circumstances to introduce a $CO_2$ tax that would place costs upon industry. At a time when environmental concerns were becoming more prominent and the economy was growing, the Finnish government did not want to give the political opposition parties an asset for the next elections and therefore was willing to strike a deal with the Greens to introduce environmental taxes in exchange for some income tax cuts.

The change of the $CO_2$ tax to base it on 25 per cent on energy content of the primary energy source and 75 per cent on the carbon content was made to take away the fiscal advantage the pure carbon tax system placed on nuclear power and imported energy. Peat as a domestic energy source should be exempt from the $CO_2$ tax to support regional and employment policy as well as for energy security reasons.

The government remained under pressure regarding its $CO_2$ and Energy tax. The 75/25 $CO_2$ tax remained subject to opposition as industry favoured an energy tax that was not based on the carbon content and the European Commission criticized the Finnish energy tax, which imposed a levy on energy imports from other Member States in the newly liberated Nordic electricity market. As the biggest power company in Finland lost clients as Danish coal power imports increased, the pressure on the $CO_2$ tax mounted. As a consequence, over time a complex compromise arose in which a series of measures were taken: the energy tax was reoriented towards an electricity consumption tax, income taxes were lowered while the electricity taxes for households (not industry) were raised and the carbon tax was removed from electricity production while heat production was taxed according to its carbon content. Moreover energy-intensive companies could now benefit from a tax refund scheme.

## 6.3 Sweden

The Swedish $CO_2$ tax was introduced at a point in time when environmental concerns were high on the social and policy agendas.[47] The

Environmental Tax Commission was set up in 1987 to analyse the possible introduction of environmental taxes in Sweden.[48] The Environmental Tax Commission was based on a broad involvement of stakeholders including politicians, bureaucrats and various interest groups and underwent a public hearing and proposed various environmental taxes including on $CO_2$, $NO_x$ and $SO_2$.

It was, however, not only the environmental mindedness of the Swedish that paved the way to the implementation of an ETR. In the late 1980s the Swedish economy was in distress and in part due to the combination of inflation and tax schedules being denominated in nominal currency, middle-income earners found themselves paying marginal income tax rates that were intended for the very rich (80 per cent marginal income taxes). As a consequence the reduction of the income tax became a policy priority and in order to reduce budget deficit increases new tax bases needed to be identified: environmental taxes such as the $CO_2$ tax were an obvious way forward.[49]

Fearing anticompetitive effects, Swedish industrial organizations opposed the $CO_2$ tax. While initially introduced without discriminating between industry and households, this changed in January 1993 when industry was exempted from all energy taxes and only had to pay 25 per cent of the $CO_2$ tax.

A benefit that the Swedish industry – and Swedish administration – enjoyed was the administrative simplicity of the introduced tax regime, which had done away with the complicated application of energy tax concessions under the pre-1993 energy tax regime.

In the case study countries similar challenges for mustering policy support for implementing $CO_2$ taxes have been encountered. The approaches to address these bear some similarities. In all countries, albeit to a varying degree, there was concern about the high income tax which was either traded in political bargaining in return for introducing a $CO_2$ tax (in the case of Finland) or where the $CO_2$ tax was used to raise funds to finance an income tax reduction (Denmark and Sweden). Income tax reduction paired with a generally heightened concern for the environment in all countries created the basis for support from households.

Industry appeared to be resisting the implementation of $CO_2$ taxes and successfully helped to shape derogation policies. These came in the form of reductions in social security contributions, energy efficiency schemes and special provisions for energy-intensive industries. In all countries the brunt of the $CO_2$ tax burden is borne by households. In Finland, where fewer derogations exist, it is noticeable that the $CO_2$ level is generally lower than in Denmark or Sweden. Besides these points administrative simplicity was also regarded as a positive element in the implementation of $CO_2$

taxes. This took the form of relying on existing tax forms or procedures or collecting relevant information for taxation at a limited number of installations.

Other policy considerations played a role in creating support for the implementation of a $CO_2$ tax. In Denmark it was the earmarking of funds for the district heating system and the desire to support the natural gas market (industrial policy) that helped the implementation. In Finland it was the pre-election expedient of not allowing the opposition party to claim this policy field that aided the introduction of the tax.

## 7 CONCLUDING REMARKS

There have been important barriers and success factors which enabled the introduction of the $CO_2$ taxes in the case study countries. The treatment above has shown that similar impediments were at play in all three countries. These impediments relate to recycling, competitiveness and the fostering of support. The delineation between these elements is not always clear cut and often there is a close interrelation between them. 'Issue linking' to strike a balance between different interests has been of paramount importance in all countries. Recycling money back to industry can improve companies' competitive position and hence appease them and foster political support or at least lead to less resistance.

The experience of the case-study countries shows that the introduction of the $CO_2$ taxes was possible by employing a consensus approach. In the case of Sweden and Denmark the $CO_2$ tax introduction was driven by a wider policy framework while in Finland it was initially a political agreement between the government and the opposition. In all countries the political resilience of the $CO_2$ taxes was ensured by frequent adaptations of either the $CO_2$ tax or its wider framework, the ETR. The consensus approach underlines the importance of recycling in the policy design and the need to safeguard competitiveness. Both issues are closely related as they can be used to keep stakeholders happy – though this should not go so far as to significantly reduce the environmental impact of the measure, as was the case in Norway.[50] In the case study countries' households received inter alia income tax reductions but were bearing a bigger share of the tax burden while companies were at least in part able to receive tax exemptions or tax refunds. Notably in Denmark cross-subsidization between households and companies was avoided. This is also a successful approach that has been followed by Switzerland, which recycles $CO_2$ tax proceeds back to residents via reductions in the health care insurance premium. In the case study countries' companies also benefited from energy efficiency schemes that were designed to help them to reduce production costs. Finland is a special case in this regard as for a long time it did not have such derogations and the Finnish $CO_2$ tax also did not benefit from flanking support of an ETR that could offer additional possibilities to support stakeholders. Perhaps this is why the Finnish tax was introduced at a relatively low rate and only increased as provisions favouring industry (e.g. in the energy domain) were extended.

It appears that industry was strongly considered and regarded as an important stakeholder while households were playing a lesser role. Perhaps this can be explained by pointing towards collective action problems that hinder households from undertaking action or the acceptance of the environmental goals as a policy justification.

In the introduction of the $CO_2$ taxes in Nordic countries we see that industrial policy goals have been an important point of consideration. This found expression in the special policies that were applicable to industry in general or energy-intensive branches in particular, which enjoyed lower tax levels or refund schemes.[51]

Perhaps as important as overcoming the barriers for introducing a $CO_2$ tax is to have a favourable policy environment for introducing it. All countries had experienced a significant degree of economic strife and used this impetus for fiscal reforms or to unlock different funding sources. Arguably one lesson to be learned is not to waste a good crisis.

## ACKNOWLEDGEMENTS

The research is part of the project ‘CATs – Carbon Taxes in Austria: Implementation Issues and Impacts’, which was funded by the Austrian ‘Klima- und Energiefonds’ and carried out within the Austrian Climate Research Programme ACRP.

## NOTES

1. This description follows Andersen, M.S. (2016).
2. Andersen, M.S. (2016).
3. The Netherlands has recently taken carbon taxation up into its recent government coalition agreement, see Regeerakkord (2017).
4. Eklund, K. (2011).
5. Berggren, H., Trägârdh, L. (2011).
6. Berggren, H., Trägârdh, L. (2011) p. 14.
7. Henriksen, I. (2006) p. 11.
8. Henriksen, I. (2006) p. 12.
9. Eklund, K. (2011) p. 9.

10. Steel, pulp and paper, shipbuilding, and mechanical engineering were in distress.
11. See Fölster, S., Kreicbergs, J. (2014) pp. 5 ff.
12. Fölster, S., Kreicbergs, J. (2014) pp. 5 ff.
13. Eklund, K. (2011) p. 9.
14. Nordic Council of Ministers (2006) p. 64.
15. Speck, S. (2007) footnote 11. See also Green Budget Europe and The Danish Ecological Council (2014) p. 3.
16. Speck, S. (2008) p. 44.
17. Speck, S. (2007) p. 34.
18. Speck, S. (2008) p. 45. For a detailed description of the tax refund scheme see Speck, S. (2007) p. 34.
19. Green Budget Europe and The Danish Ecological Council (2014) p. 3.
20. Speck, S. (2008) p. 46.
21. Speck, S. (2007) p. 38.
22. Speck, S. (2007) p. 39.
23. Speck, S., Jilkova, J. (2009) p. 32.
24. Speck, S., Jilkova, J. (2009) p. 33.
25. Vehmas, J. (2005) p. 2181.
26. Speck, S. (2008) p. 53.
27. Speck, S. (2008) p. 50.
28. Speck, S. (2008) p. 44 and Speck S. (2007) p. 34, footnote 11.
29. Speck, S. (2007) p. 34, footnote 12.
30. Speck, S. (2008) p. 36.
31. Speck, S. (2007) p. 36.
32. Speck, S. (2007) p. 71.
33. Speck, S. (2007) p. 38.
34. Light processes enjoyed a reduction of around 24% while heavy processes enjoyed initially 40% reduction (in 1996). As the $CO_2$ tax was raised from 5 DKK in 1996 through the years to 25 DKK per ton of $CO_2$ in 2000, also the percentage of the reduced tax rate increased to 88% because the tax rate for heavy processes under the government agreement remained fixed at 3 DDK per ton of $CO_2$. See Speck S. (2007), table A4-1c; and see also Nordic Council of Ministers (2006), p. 64.
35. Vehmas J. (2005) p. 2180.
36. Vehmas, J. (2005) pp. 2177–8.
37. Vehmas, J. (2002).
38. Vehmas, J. (2002) p. 250.
39. Vehmas, J. (2005) p. 2177 and Sairinen, R. (2012) p. 431, footnote 4.
40. Johansson, B. (2006) p. 2 and Hammar, H., Åkerfeldt, S. (2011).
41. Speck, S. (2007) section 4.6.2.
42. Sterner, T. (1994) p. 22.
43. Speck, S. (2007) section 4.6.2.
44. Speck, S. (2007) p. 34.
45. Speck, S. (2007) p. 35.
46. This section is based on Sairinen, R. (2012) pp. 426 ff.
47. Hammar, H., Åkerfeldt, S. (2011).
48. Sterner, T. (1994).
49. This passage follows Sterner, T. (1994).
50. Bruvoll, A., Larsen, B.M. (2004).
51. Vehmas, J. (2005) p. 2176.

## REFERENCES

Andersen, M.S. (2016) An Introductory Note on Carbon Taxation in Europe, A Vermont Briefing, paper presented at Carbon Pollution Taxes: A Conversation with International Experts, Vermont College of Fine Arts, Montpelier 12/01/2016, available at: https://www.researchgate.net/publication/313023514_An_Introductory_Note_on_Carbon_Taxation_in_Europe_-_A_Vermont_Briefing

Berggren, H. and Trägårdh, L. (2011) Social Trust and Radical Individualism – The Paradox at the Heart of the Nordic Capitalism, World Economic Forum Davos 2011

Bruvoll, A. and Larsen, B.M. (2004) Greenhouse Gas Emissions in Norway: Do Carbon Taxes Work? Energy Policy, vol. 32 (2004), 493–505

Eklund, K. (2011) Nordic Capitalism: Lessons Learned in The Nordic Way, Shared Norms for the New Reality, World Economic Forum Davos 2011

Fölster, S. and Kreicbergs, J. (2014) Twenty Five Years of Swedish Reforms, Reforminstitutet

Green Budget Europe and The Danish Ecological Council (2014) Successful Environmental Taxes in Denmark, https://green-budget.eu/wp-content/uploads/The-most-successful-environmental-taxes-in-Denmark-2_FINAL.pdf

Hammar, H. and Åkerfeldt, S. (2011) $CO_2$ Taxation in Sweden – 20 Years of Experience and Looking Ahead

Henriksen, I. (2006) An Economic History of Denmark. EH.Net Encyclopedia, edited by Robert Whaples. 6 October. http://eh.net/encyclopedia/an-economic-history-of-denmark/

Johansson, B. (2006) Economic Instruments in Practice 1: Carbon Tax in Sweden, Swedish Environmental Protection Agency, Working Paper (Stockholm: Swedish Environmental Protection Agency)

Nordic Council of Ministers (2006) The Use of Economic Instruments in Nordic and Baltic Environmental Policy 2002–2005, Copenhagen, Denmark

Regeerakkord (2017) Vertrouwen in de toekomst, Regeerakkoord 2017–2021, VVD, CDA, D66 en ChristenUnie

Sairinen R. (2012) Regulatory Reform and Development of Environmental Taxation: The Case of Carbon Taxation and Ecological Tax Reform in Finland, in Janet Milne and Mikael Skou Andersen (eds), Handbook of Research on Environmental Taxation, Edward Elgar Publishing, pp. 422–38

Speck S. (2007) Overview of Environmental Tax Reforms in EU Member States, WP1, Competitiveness Effects of Environmental Tax Reforms (COMETR), Final report

Speck, S. (2008) The Design of Carbon Broad-based Energy Taxes in European Countries, Vermont Journal of Environmental Law, vol. 10, 31–59

Speck, S. and Jilkova, J. (2009), Design of Environmental Tax Reforms in Europe, in Mikael Skou Andersen and Paul Ekins (eds), Carbon-Energy Taxation: Lessons from Europe, Oxford University Press

Sterner, T. (1994) Environmental Tax Reform: The Swedish Experience, Environmental Policy and Governance, vol. 4, issue 6, 20–25

Vehmas, J. (2002) Money for Sweden and $CO_2$ Emissions to Denmark – Reconstitution of the Finnish Environment-based Energy Taxation in 1993–

1996, Acta Universitatis Tamperensis 861 (in Finnish, English summary), available at: http://tampub.uta.fi/handle/10024/67192
Vehmas, J. (2005) Energy-related Taxation as an Environmental Policy Tool – the Finnish Experience 1990–2003, Energy Policy, vol. 33, 2175–82

# 2. Border adjustment with taxes or allowances to level the price of carbon

**Mikael Skou Andersen***

## 1 INTRODUCTION

The adoption of the Paris Agreement on reduction of greenhouse gas emissions was a landmark decision, as it codified a broad preparedness to prevent global warming from exceeding a 2 degree target through concerted action among the nations of the world in the context of the United Nations Framework Convention on Climate Change (UNFCCC 2015). In contrast to the Kyoto Protocol, which established specific targets mainly for industrialised countries, the Paris Agreement refers target setting to the signatories and includes all nations. Rather than a top-down framework of firm commitments, the architecture of the Paris Agreement provides for a bottom-up approach, where countries will report on a regular basis their nationally determined contributions (NDCs) towards emission reductions.

Notwithstanding its voluntary approach to target setting, the Paris Agreement maintains that reductions should be rapid so as to enable emissions to peak as soon as possible (art. 4.1), and it makes a call for ambitious efforts (art. 3). Questions about the choice of policy instruments to achieve NDCs will be the prerogative of Parties to the agreement. Still, countries must report their emissions along with information that is necessary to track progress made in implementing and achieving NDCs (art. 13.7). In this context a subsidiary body, the 'Forum on the impact of the implementation measures', will serve the agreement by analysing impacts of mitigation actions and promoting the exchange of information, experiences and best practices among parties (III.34).

Moreover, a new coalition of actors willing to promote carbon pricing as a key tool in emissions reductions emerged from the 21st Conference of the Parties to the UNFCCC in Paris. A 'Carbon Pricing Leadership Coalition' (CPLC), launched by the World Bank and the International

Monetary Fund, was founded with the aim to increase the coverage of greenhouse gas emissions by carbon pricing tools two-fold by 2020 and four-fold by 2030, to reach 50 per cent coverage of global emissions.[1] The CPLC includes countries, international institutions and business firms. The 25 founding government partners have a notable share in global emissions (e.g. California, Canada, France, Germany, Mexico, Spain and the UK). In the absence of explicit decision making on carbon pricing in the context of the UNFCCC, the CPLC will hopefully be able to help carbon pricing gain momentum as a key measure to curb emissions and create markets for low-carbon technologies.

The Intergovernmental Panel on Climate Change (IPCC) in its fourth assessment report published 10 years ago warned that a carbon price would need to reach €15–20/t$CO_2$ to provide sufficient incentives for a low-carbon development that could lead to stabilisation of atmospheric concentrations of $CO_2$ at a level of 550 ppm (IPCC 2007). However, for a more stringent stabilisation scenario below 450 ppm, as implied by a 2 degree target, the IPCC estimated that the implicit carbon price required would probably be closer to €30–40 per ton $CO_2$ from 2020. The IPCC experts suggested that in any case the carbon price would need to increase annually by €1–2 per ton $CO_2$ in real terms – indefinitely – to enable the long-term reduction of emissions required for stabilisation within 2 degrees of warming. In this context it might be appropriate to recall that if the 2 degree target is not attained, there are likely to be considerable climate change challenges in a future world, with long-term sea level rises above present levels predicted by ice-core research to 4–8 metres (comparable to conditions in the Eem period 120,000 years ago).

## 2 CARBON PRICING – STATE OF THE ART AND RECENT DEVELOPMENTS

From about 1990 and for more than a decade, four Nordic countries were pioneering the use of carbon taxes. Towards the end of the 1990s they were joined by three smaller transition economies (Slovenia, Estonia and Latvia) (Andersen 2018).

Finally, by 2005 the EU introduced its emissions trading system (ETS) with allowances for carbon emissions (2003/87/EC). Initially, all allowances were handed out for free ('grandfathered'), but from the inception of phase 2 in 2007 partial auctioning of allowances was introduced. Prior to the 2008 financial crisis, trading saw a gradual increase in the carbon price up to a level of €22/t$CO_2$, but since then the price has levelled off.

The ETS applies to designated large installations, including power and heat production. A few European countries have enacted complementary taxes to price carbon for the non-ETS sectors. In addition to the above-mentioned countries, this applies for France, Ireland and Portugal, while the UK has a climate change levy as well as a price floor for ETS-covered installations. Switzerland has joined the ETS and introduced a carbon tax too (Kossoy, Klein et al. 2015).

By 2012, two large non-European economies, Australia and Japan, enacted a carbon price signal. The rate of Japan's carbon tax represents with about €2/t$CO_2$ a very cautious beginning (Lee et al. 2012). Australia, on the other hand, constrained its carbon price mechanism (tax transforming into an ETS) to large industrial installations, but experienced a sudden repeal following a change of government in 2015 (Meng et al. 2013).

In North America carbon pricing developed along two different routes from about 2008. A group of northeastern US states created a carbon allowance system applying to the power sector (RGGI), while, in Canada, three provinces opted for carbon taxes with wider coverage of the economy. The province of British Columbia is alone in having a carbon price matching levels in Europe (Rivers and Schaufele 2012).

In the past five years, carbon pricing has gained some momentum in a number of countries outside the set of industrialised countries listed in annex 1 to the former Kyoto Protocol, so that different schemes are emerging in countries as diverse as Mexico, Korea, Chile, South Africa, New Zealand, Kazakhstan and China. As a result of these efforts, the share of global $CO_2$ emissions subject to a price has tripled from 4 per cent in 2011 to more than 12 per cent presently. Most of these countries are beginning from very modest tax rates in line with Japan, but identification of carbon as a tax base may allow for more ambitious policies to come (Kossoy, Klein et al. 2015).

China is of particular interest. It has created an emissions trading scheme effective from 2013 in seven pilot areas, including its major cities Beijing, Shanghai and Guangdong, where allowances are traded at prices of up to $7/t$CO_2$. By 2018 its carbon emissions trading scheme is expected to be covering large installations in the entire country. Concurrently, China's 12th five-year plan mandates the introduction of a carbon tax, presumably for non-trading sectors. The framework for these measures is an ambition to limit emissions by 2030 with up to two-thirds per unit of GDP (Lo 2015).

Finally, from 2013 an emissions trading scheme became effective in the state of California. This may serve as a pilot for a comprehensive scheme in the USA, as it includes not only the power sector, but also large industrial installations and, from 2014, fuel distributors for transport and heat, whereby about 85 per cent of the state's greenhouse gases are covered.

Under this scheme, allowances are traded at prices of about \$12/t$CO_2$, while about 10 per cent are auctioned and the remaining 'grandfathered'.[2]

## 3 PROSPECTIVE BORDER ARRANGEMENTS FOR CARBON PRICING

The emergence of carbon pricing has triggered interest in finding methods whereby it can be avoided that businesses resident outside economies with carbon pricing can gain a competitive advantage that would be harmful and lead to carbon leakage.

The European Parliament (2014), in a resolution on the steel sector, called on the European Commission 'to tackle, in a timely and effective manner, steel imports into the EU market which have been illegally subsidised and dumped, and to use, where appropriate, the EU trade remedy instruments in line with existing EU law'. In this context the European Parliament specifically requested the European Commission 'to examine the feasibility of a border carbon adjustment (payment of ETS allowances for steel coming from outside the EU) with a view to creating a level playing field in terms of $CO_2$ emissions, thus eliminating the phenomenon of carbon leakage'.

The universal principle behind border adjustment is simple: tax rates are adjusted at the border, so that imports from non-carbon-priced foreign competitors are taxed equivalently to the domestic carbon price. At the same time, exports may be rebated by the amount of carbon price they bear, when sold to markets without carbon pricing. This enables a country to price carbon for its domestic industry for the purposes of meeting its NDCs while preserving its competitiveness abroad by allowing its exports to compete in markets without carbon pricing, and taxing imports to the same degree (Brack 1998).

Where carbon is priced through an ETS, a border adjustment may function in a way comparable to a tax in that foreign producers may be requested to hand over a number of emissions allowances equivalent to the emissions of their products. Conversely, exports to non-carbon-pricing markets receive a refund of the associated carbon emissions allowances. In the context of preparing the next phase of the EU ETS, the European Parliament's Environment Committee submitted a proposal to the plenary to instate a border adjustment for the cement industry, while ending the grandfathering scheme for this sector (Mehling et al. 2017, p. 28).

Border adjustments are of interest as they are applied in many other areas. Tax rates for gas and diesel are the same for domestic and imported fuels, and do not apply where motor vehicle fuel is exported. Cigarettes and alcohol are other areas with an established tradition of border adjustments with respect to tax rates.

When considering border adjustment with respect to carbon, the situation is more complicated. It is not fuels per se that are causing complications. Rather, the main practical and legal challenges concern products that have been manufactured and processed with the *use* of fossil fuels. Producers in non-carbon pricing (non-CP) markets would be required to declare the amount of fossil fuel carbon embodied in the products for sale into carbon price (CP) markets. It is more straightforward when a refund is to be offered for export of products from CP markets into non-CP markets, as the information would already be available from established systems of emissions accounting (national and/or to UNFCCC).

GATT article II.2(a) allows WTO (World Trade Organization) members to impose on the import of any product a charge equivalent to a domestic tax. At the same time basic WTO principles prescribe non-discrimination according to national origin and between like products, which may create concerns (Tamiotti, Olhoff et al. 2009).

It has been suggested that the GATT agreement should confine the option of introducing a border arrangement to products per se, while not allowing the related production *processes*, including the specific energy uses, to be taken into account. This would impair any border arrangement for imports based on carbon emissions, since these are not embodied in the product. Still, under GATT article III.2, border adjustments on imported products are allowed in respect of taxes 'applied, directly or indirectly, to like domestic products'. It has been argued that the term *indirectly* allows for border adjustments relative to the inputs used during the production processes, e.g. carbon-based fuels. This view is supported by GATT's panel ruling in the *Superfund* case, where a tax on substances used as input for the production of chemicals was found eligible for border tax adjustment. The US tax on Ozone Depleting Substances is a further example, where border tax arrangements could apply to products containing and produced with such substances, subject to a minimum threshold.

As to possible tax refunds for exported products, that would be part of a border arrangement, these are permissible under WTO rules as long as they do not exceed the domestic excise taxes accrued, and hence the refunds cannot be considered a subsidy. Such export border arrangements are not prohibited nor can they become subject to countervailing measures. There seems to be general agreement that a domestic tax on fuels can be rebated when a product is exported.

The GATT and WTO agreements have some other general provisions that are of interest too. Besides the national treatment principle (GATT art. III.2), the most-favoured nation clause (GATT art. I.1) is also of key interest. Essentially both are non-discrimination principles. The national treatment principle requires that imported goods are not treated differently from domestic goods, while the most-favoured nation clause requires that all WTO members are treated in the same way with respect to their traded products.

Measures found inconsistent with core provisions of GATT could still qualify for exemptions from the general set of rules. GATT article XX details the exemptions. According to subparagraphs (b) and (g) of GATT art. XX, policy measures that are necessary 'to protect human, animal or plant life or health' (b) or 'relating to the conservation of exhaustible natural resources' (g) could provide justification. However, a WTO member must demonstrate that a policy measure is not an arbitrary discrimination or simply a disguised restriction on international trade (Olsen 2012).

In the past, several measures have been considered by dispute panels, so there is a rich case law relating to these two subparagraphs of art. XX. For instance, measures aiming at the prevention of air pollution and the depletion of clean air have been found legitimate in qualifying for exemptions.[3] In the *Shrimp-Turtle* dispute, the Appellate Body decision allowed for use of the exemption clause, once the most-favoured nation principle had been secured, whereby the essentially process-based trade restrictions (shrimp fishing methods endangering turtles) were not an issue. Legal counsellors at the WTO have interpreted this ruling to suggest that border arrangements for carbon pricing could be accepted as legitimate under the exemption clauses, when all other options have been exhausted (Tamiotti, Olhoff et al. 2009, p. 108, note 272).

A border arrangement will require carbon content of products to be estimated. While European firms are obliged to give information on their emissions under the EU ETS (art. 14), it might prove difficult to impose similar requirements on non-EU firms trading with the EU. One solution would be to make declaration of carbon emissions voluntary, while applying the product-specific benchmarks established under the EU ETS for industry products in the absence of any declarations (Monjon and Quirion 2011; Commission Decision 2011/278/EU).[4] As these benchmarks refer to the best-performing companies, it cannot be claimed that they will discriminate against non-EU producers. The drawback is that they might actually underestimate the embodied emissions. For energy-intensive products, such as iron and steel, cement and aluminium, it might be better to introduce a compulsory, third-party verification scheme for emissions. As imports would come from a limited number of companies, it seems to be administratively feasible. The approach in proposal #84 on the cement sector for the EU ETS was to make reference to the *average* emissions of

EU producers, minus any benchmark-based free allocation. In the end this proposal did not win approval in the European Parliament, but in particular France continues to explore the options (Mehling et al. 2017).

## 4 THE FUTURE OF CARBON PRICING

The EU ETS has not been living up to the high expectations of policymakers as the price for carbon allowances over the past years has been well below the level required to provide incentives for a low-carbon development (Zapfel 2005). Allowances have been traded at prices of €5–8 per ton of $CO_2$ since the financial crisis of 2008 and there is a huge reserve of allowances in the market. Market observers point to the over-allocation of free allowances to ETS installations outside the electricity sector, essentially comprising all the large energy-intensive industries, of which iron and steel, cement, aluminium, chemicals, refineries, and pulp and paper are the most important (Carbon Market Watch 2017). These six industries account for the greater share of industry's carbon emissions, while their contribution to GDP represents only about 2 per cent.

In the end, the EU was able to meet its 8 per cent reduction target under the Kyoto Protocol, but the ETS played a fairly small role in this achievement. Studies suggest that the introduction of the ETS helped curb emissions by about 2–3 per cent relative to business as usual in the very first phase and prior to the financial crisis (DECC 2012, p. 18). Developments after 2008 are hard to disentangle because of volatility in the markets, but the economic downturn left many companies with a large reserve of allowances that could be banked for future use, and which depressed the price the allowances could be traded at. With allowances handed out for free to industries in the first and second phases of the ETS, these emission permits may in fact have enabled the continued operation of sunset industries, as firms could sell surplus allowances and from the proceeds continue to finance their operations (de Bruyn et al. 2010).

As a result of the free-allocation approach, potential revenues have been lost for governments otherwise struggling with budget deficits. Some observers maintain that energy-intensive industries have been able to pass the value of the allowances into their product prices, whereby consumers have experienced increased commodity prices. An evidence review published by the UK's Department of Energy and Climate Change (DECC 2012, p. 5) noted that '[t]here was fairly robust evidence based on price data that a number of sectors were able to pass through the costs of emission permits on to final product markets'. A previously published report presented advanced econometric analysis of three energy-intensive sectors and came to a similar conclusion (de Bruyn et al. 2010). The free allocation was granted in the belief that energy-intensive sectors could not pass through the costs of auctions into the product prices. The main effect is then that companies have received a windfall profit from the EU ETS and that income from citizens has been transferred to companies. From an economic perspective, it seems quite straightforward that companies will pass the opportunity costs of allowances into their product prices when they have the market power to do so. Nevertheless, in the negotiations over the next and fourth phase of the ETS, which will begin from 2020, energy-intensive industries have been making their voice heard in favour of continued free allocation, and they have required compensation for the costs of climate mitigation policy.[5] Some legal experts contend that support to energy-intensive industry with free allowances should be notified as state aid subsidies to the WTO (Aydos 2017).

It is worthwhile considering a carbon border arrangement as an alternative to continued free allocation, as a border arrangement will probably provide better protection against carbon leakage and unfair competition from companies in non-CP countries without creating huge windfall profit distortions in the commodity markets. Following the initial proposal of former French President Nicolas Sarkozy to introduce a carbon border arrangement, studies have been looking into the consequences for emissions, economic performance and competitiveness.[6]

Monjon and Quirion (2011) explored the consequences of a border arrangement while considering several different scenarios for their architecture, based on a tax as well as allowances. These scenarios reflect some of the options reflected above, specifically whether use should be made of an EU benchmark for the emissions of foreign producers or whether, alternatively, a proxy for their actual emissions in the rest of the world should be used. The modelling and analysis also considered the implications of moving to full auctioning, as well as restricting the border arrangements to imports, whereby no refunds would be offered for exports out of the EU. The analysis focuses on sectors responsible for 75 per cent of emissions from ETS installations, covering steel, cement, aluminium and electricity.

Their findings show that the revenues from a carbon price of about €20/t$CO_2$ would amount to somewhere between €20 and 30 billion (Monjon and Quirion 2011). The border arrangement on its own would generate up to €2 billion in revenues, most when a proxy for emissions in the rest of the world is used for the imports from non-CP countries. A key finding is that a border arrangement will reduce world carbon emissions more than a system with free allocation. This provides a clear environmental justification, which is important for WTO compatibility under art. XX of the GATT. They find that an allowance-based border arrangement performs somewhat better than one based on a tax as it will be more effective in limiting carbon leakage. Another interesting finding is that the decrease in global emissions is larger when exports are included in the border arrangements; this is due to EU commodities being less carbon intensive than those produced in the rest of the world. Still, the EU would apparently not win market shares to an extent that could make claims over disguised trade restrictions under the border arrangement credible. This is because EU market shares would decline compared to a free-allocation system. Production will decrease in EU industries under all scenarios, but least with the system of free allocation, which may add a further argument for preserving it in the view of energy-intensive industries. There will be a decrease in steel and cement production as demand will be curbed when the costs of carbon are factored in. Sectors that make use of steel and cement, such as construction and automotive production, will experience a knock-on effect, increasing their costs and reducing demand for their products relative to other commodities. This would be an efficient outcome from a scheme that diffuses the carbon price signal into the economy, leading to reductions in emissions where their marginal value is the least.

The analysis of Monjon and Quirion hints that revenues from full auctioning and border adjustment could be used to consolidate government budgets as is much needed in several EU member states. Alternatively, revenues could be recycled to lower other taxes. Traditionally, carbon tax revenues have been used to lower payroll taxes for employers and employees, thereby providing a stimulus to labour-intensive sectors of the economy. Their analysis does not explore what impacts would follow from the stimulus side of the equation of revenue generation, which is a separate and complex question relating to the opportunity for a possible 'double dividend' (see Andersen and Ekins 2009).

## 5 FINAL REMARKS: THE DIPLOMACY OF BORDER CARBON ADJUSTMENT

When confronted with global warming, the advantages of opting for a price on greenhouse gas emissions are clear. However, carbon pricing has proven to be a contentious issue in every part of the world. Constituencies will inevitably ask why they should pay for their emissions, when others are free to continue polluting, while questions over impacts on competitiveness are blurring policy making. The uptake of carbon pricing has been slow, with the reluctance to do so reflecting a half-hearted affair with widespread exemptions and handing out of free allowances, causing unanticipated side-effects and complications economically, legally and politically.

In some respects, the Paris Agreement is a watershed in the history of climate change mitigation. It provides an end to purely unilateral action as from now on all State Parties to the agreement must contribute to curbing greenhouse gas emissions. Developments over the past few years have shown that more countries will include some amount of carbon pricing in their schemes of policies and measures, and they will be supported by economic analysts endorsing market-based policy instruments as the most cost-effective mitigation approach. Yet, as countries report their emissions and measures to the UNFCCC, it will become clear within the next decade whether there are also gaps and free riders. If so there will inevitably be a discussion as to how to level the playing field.

Many trade specialists have expressed profound scepticism about border arrangements for carbon, pointing to the complex legal arrangements under international trade law. Yet, it cannot be ruled out that an arrangement could win approval under GATT art. XX's exemption clauses (Tamiotti 2011; 2014). The practices of the WTO suggest that a border arrangement will be approved only if all other measures have been exhausted and there have been prior international negotiations among the concerned parties to explore the options (Branger and Quirion 2014).

It is essential to see border arrangements as a diplomatic tool, perhaps even more so than an actual tool. With respect to aviation, the EU passed a directive that imposed on flights from outside Europe an obligation to accede to the ETS for carbon allowances. In practice, this obligation was comparable to a border arrangement for carbon. The hefty and negative reaction from all the major countries affected, including the USA, China and India, forced the EU to suspend the arrangement while a deal on aviation emissions was negotiated. Notwithstanding the controversies, the recent agreement at the assembly of the International Civil Aviation Organisation (ICAO) to introduce a global scheme for offsetting emissions was achieved in the shadow of a viable border arrangement. Sixty-six countries representing 87 per cent of aviation activity have agreed to participate in the scheme, while allowing certain low-income, developing countries exemptions.[7]

Border carbon adjustment has been a recurrent theme in the USA's concerns over rapidly rising emissions from China and was explicitly mentioned in the Waxman-Markey proposal on an allowance system with border arrangements. This caused much anger in Beijing in the run-up to the 2009 climate conference in Copenhagen. China voiced concerns about green protectionism and even voiced, in the 2012 Rio Summit, reservations and suspicions about any references to the green economy as possible euphemisms for trade restrictions. Still, it also stimulated consideration over carbon pricing domestically in China and helped the climate mitigation protagonists win responsiveness in the economic and trade-oriented administrations. This led eventually to the agreement on carbon pricing in the context of the 12th five-year plan.

With trade agreements between Europe and the USA in a critical phase, and with an increasingly unpredictable outcome, there is a certain risk that indications of border carbon adjustment could cause relationships to deteriorate further or become part of a difficult climate between major trading partners. Hence, border adjustment for carbon is as much a tool in the diplomatic arsenal and needs to be mobilized in a considerate way.

## NOTES

* This chapter updates and reproduces a previously published essay entitled 'The Missing Link in an International Framework for Carbon Pricing: Border Adjustment with Taxes or Allowances', in M. Villar-Ezcurra (ed) (2017), *State aids, taxation and the energy sector*, Madrid: Thomson-Reuters Aranzadi.

1. https://www.carbonpricingleadership.org/, accessed 1 Feb 2018.
2. https://www.c2es.org/content/california-cap-and-trade/, accessed 1 Feb 2018.
3. US – Gasoline: Panel Report, United States – Standards for Reformulated and Conventional Gasoline, WT/DS2/R, adopted 20 May 1996, as modified by Appellate Body Report, WT/DS2/ AB/R, DSR 1996:I, 29.
4. http://eur-lex.europa.eu/legal-content/EN/TXT/PDF/?uri=CELEX:32011D0278&from=EN, accessed 1 Feb 2018.
5. https://www.euractiv.com/section/energy/news/oil-refiners-warn-increased-co2-costs-could-force-them-out-of-europe/, accessed 1 Feb 2018.
6. https://www.euractiv.com/section/climate-environment/news/sarkozy-renews-pressure-for-co2-border-tax/, accessed 1 Feb 2018.
7. https://icao.int/environmental-protection/Pages/market-based-measures.aspx, accessed 1 Feb 2018.

## REFERENCES

Andersen, Mikael Skou (2018, in press), 'The politics of carbon taxation in Europe', in Roberta Mann and Tracey Roberts (eds), *Tax and the environment*, Lanham: Lexington Books.

Andersen, Mikael Skou and Paul Ekins (eds) (2009), *Carbon-energy taxation: Lessons from Europe*, Oxford: Oxford University Press.

Aydos, Elena de Lemos Pinto (2017), *Paying the carbon price: The subsidisation of heavy polluters under emissions trading schemes*, Cheltenham: Edward Elgar Publishing.

Brack, Duncan (1998), 'Energy/carbon taxation, competitiveness & trade', Energy Economist: an international analysis, *Financial Times Energy Newsletters* 206, p. 7.

Branger, Frédéric and Philippe Quirion (2014), 'Climate policy and the "carbon haven" effect', *WIREs Climate Change*, 5(1), 53–71.

de Bruyn, Sander, Agnieszka Markowska, Femke de Jong, Mart Bles et al. (2010), *Does the energy intensive industry obtain windfall profits through the EU ETS? An econometric analysis for products from the refineries, iron and steel and chemical sectors*, Delft: CE Delft.

Carbon Market Watch (2017), *A fair EU ETS revision*, Policy Briefing, Brussels.

DECC (Department of Energy and Climate Change) (2012), *An evidence review of the EU emissions trading system, focusing on effectiveness of the system in driving industrial abatement*, London: DECC.

European Parliament (2014), Resolution of 17 December 2014 on the steel sector in the EU: protecting workers and industries (2976(RSP)), Texts adopted P8_TA(2014)0104.

IPCC (Intergovernmental Panel on Climate Change) (2007), *Fourth Assessment Report, Mitigation of climate change, Working Group III*, Cambridge University Press.

Kossoy, Alexandre, Noémie Klein et al. (2015), *State and trends of carbon pricing*, Washington DC: World Bank Group and ECOFYS.

Lee, Soocheol, Hector Pollitt and Kazuhiro Ueta (2012), 'An assessment of Japanese carbon tax reform using the E3MG econometric model', *The Scientific World Journal*, Article ID 835917.

Lo, Alex (2015), 'Too big to fail: China pledges to set up landmark emissions trading scheme', *The Conversation*, 26 Sep, http://theconversation.com/too-big-to-fail-china-pledges-to-set-up-landmark-emissions-trading-scheme-48214.

Mehling, Michael, Harro van Asselt, Kasturi Das, Susanne Droege and Cleo Verkuijl (2017), *Designing border carbon adjustments for climate action*, London: Climate Strategies.

Meng, Sam, Mahinda Siriwardana and Judith McNeill (2013), 'The environmental and economic impact of the carbon tax in Australia', *Environmental and Resource Economics*, 54(3), 313–32.

Monjon, Stéphanie and Philippe Quirion (2011), 'A border adjustment for the EU ETS: Reconciling WTO rules and capacity to tackle carbon leakage', *Climate Policy*, 11(5), 1212–25.

Olsen, Birgitte Egelund (2012), 'Gaining intergovernmental acceptance: legal rules protecting trade', in Janet E. Milne and Mikael Skou Andersen (eds), *Handbook of research on environmental taxation*, Cheltenham: Edward Elgar Publishing, pp. 192–210.

Rivers, Nicholas and Brandon Schaufele (2012), 'Carbon tax salience and gasoline demand' Working Paper #1211E, Ottawa: Dept. of Economics, University of Ottawa.

Tamiotti, Ludivine (2011), 'The legal interface between carbon border measures and trade rules', *Climate Policy*, 11(5), 1201–11.

Tamiotti, Ludivine (2014), 'The trade and climate change debate and the topic of border tax adjustments', Keynote at 15th Global Conference on Environmental Taxation, http://conferences.au.dk/fileadmin/conferences/gcet/PowerPointPresentations/Keynote_Tamiotti_GCET15.pdf.

Tamiotti, Ludivine, Anne Olhoff et al. (2009), *Trade and climate change. A report by the United Nations Environment Programme and the World Trade Organization*, Geneva: WTO & UNEP.

UNFCCC (United Nations Framework Convention on Climate Change) (2015),

Conference of the Parties Twenty-first session, *Adoption of the Paris Agreement*, 12 Dec.

Zapfel, Peter (2005), 'Greenhouse gas emissions trading in the EU: building the world's largest cap-and-trade scheme', in B. Hansjürgens (ed), *Emissions trading for climate policy*, Cambridge University Press.

# 3. Towards bottom-up carbon pricing in Canada

**Takeshi Kawakatsu and Sven Rudolph***

## 1 INTRODUCTION

While the Paris Agreement can certainly be considered a major diplomatic success, it urgently needs to be underpinned by ambitious policies in order to achieve its major target of '[h]olding the increase in the global average temperature to well below 2°C' (UN 2015, Art. 1) because current national proposals fall significantly short of this goal (UNFCCC 2016). Domestic carbon pricing is a promising way of substantiating the Paris Agreement. Economists have almost unanimously supported the use of market-based approaches to environmental problems on the grounds of their economic efficiency and environmental effectiveness (Endres 2011). It also has been shown that a sustainable design of pricing schemes such as greenhouse gas (GHG) cap-and-trade or taxes, considering not only environmental and economic, but also social justice criteria, is possible (Rudolph et al. 2012; 2014), and thus answering to the call of the Paris Agreement to also 'reflect equity' (UN 2015, Art. 3). In addition, recently carbon pricing has become more widespread, extending not only across several continents, but across all governance levels (World Bank/Ecofys 2016).

Despite these merits, ambitious market-based climate policy still faces serious implementation barriers. The Public Choice literature holds on to its skepticism towards the feasibility of ambitious GHG taxes or cap-and-trade schemes (Schneider et al. 2015). Yet, case studies have also shown political windows of opportunity (Rudolph 2005). Most recent developments in GHG pricing indicate a trend towards regional or local programs, especially in countries where national level carbon pricing has failed.

Sub-national policies are an important supplement to global and national measures. The New Environmental Federalism (Oates 2004) objects to earlier warnings of a 'race to the bottom' (Stewart 1977) and underscores the value of sub-national regimes as policy laboratories. Also, on the local and regional level policy measures can be tailored to residents' preferences and to the particular infrastructural needs in order to increase economic welfare. And Ostrom (2009) emphasized the importance of polycentric climate policy. In practice, Canada, amongst the global top ten emitters of per capita and total GHG, is an obvious example of the political failure of national carbon pricing and the success of sub-national action. As a consequence, since October 2016, Canada's Prime Minister Trudeau has tried to utilize sub-national dynamics for establishing a national carbon price.

These trends in Canadian GHG pricing – similar also to pre-Trump US trends – raise the following two questions. Can carbon pricing in Canada be considered sustainable? And what are the interdependencies between provincial and federal action? In order to answer these questions, we analyze the BC carbon tax in a case study design, but also reflect on current cap-and-trade schemes in Québec and Ontario and the 2016 Trudeau Initiative. Against the background of sustainability economics, Public Choice, and Environmental Federalism, we describe program design, analyze the results, and examine the interdependent political process.

## 2 LEADING THE WAY: BRITISH COLUMBIA CARBON TAX

### 2.1 Design and Results

British Columbia launched its 'Climate Action Plan' in 2008 with aggressive GHG targets of a 33 percent reduction from 2007 levels by 2020 (10 percent below 1990 levels) and an 80 percent reduction by 2050. In order to reduce GHG emissions, at the same time, a carbon tax was included as a core component of the province's climate change strategy. The BC carbon tax was implemented on July 1, 2008, and relied on a highly ambitious design.

The tax base includes all fossil fuels purchased for combustion within the province. However, the tax does not apply to non-combustion emissions from industrial process, landfills, forestry and agriculture. Also fuels exported from BC and fuels used for inter-jurisdictional commercial marine and aviation purposes are excluded. Coverage, hence, is 75 percent of total provincial GHG emissions.

The tax rate started at 10 C$ per ton of $CO_2$-equivalent ($CO_2e$) emissions in 2008, increased by 5 C$/t each year for the next four years in order to reach 30 C$/t in 2012, and has remained at that level since. Applying to all fossil fuels, the specific tax rates vary for each type of fuel depending on the amount of $CO_2e$ released in combustion. The carbon tax built upon existing fuel tax collection mechanisms that require tax to be collected and remitted at the wholesale level.

Every cent generated by BC carbon tax is returned to taxpayers through tax reductions and credits. Thus, it is 100 percent revenue neutral and represents a tax shift rather than a tax increase. Expenditures outlined in an annual plan can be differentiated between household tax measures and business tax measures. The former includes measures such as the Low Income Climate Action Tax Credit, a 5 percent tax reduction in the first two personal income tax brackets, and a Northern and Rural Homeowner Benefit. The latter includes, for instance, the reductions in each of general and small business corporate income tax rates and the Scientific Research and Experimental Development Tax Credit.

The BC carbon tax resulted in a significant decrease of fuel consumption and GHG emissions, while it was just introduced when the severe global recession hit in 2008/2009. Dramatic differences in fuel sales and GHG emissions between BC and the rest of Canada show that the recession alone cannot be blamed for the decline in fuel consumption in BC relative to the rest of Canada in the years following the introduction of the tax (Pedersen and Elgie 2015). From 2008 to 2012/2013, BC's fuel consumption per capita declined by 16.1 percent, while it increased by 3 percent in the rest of Canada during the same period (Elgie and Mcclay 2013). BC's per capita GHG emissions associated with carbon-taxed fuels also declined by 10 percent from 2008 to 2011 and outpaced those in the rest of Canada by almost 9 percent. Moreover, using econometric methods, Rivers and Schaufele (2015a) estimate that at \$30/t $CO_2$, the carbon tax caused a reduction of 11–17 percent in gasoline sales from 2008 to 2011. Gulati and Gholami (2015) also find that the carbon tax likely reduced natural gas consumption by 15 percent in the residential and 67 percent in the commercial sectors respectively. Based on these results, Murray and Rivers (2015) claim that the effect of the tax was to reduce fuel consumption and GHG emissions by 5–15 percent in BC.

Regarding economic effects of the BC carbon tax, the empirical evidence suggests little net impact. Elgie and Mcclay (2013) show that after four years BC's GDP has slightly (0.1 percent in total 2008–11) outperformed the rest of the country over the period the carbon tax has been in place. And this is even true despite economic constraints, including lagging demand for BC lumber from the sluggish US housing market. Considering that BC has by far the highest carbon price in North America, the data clearly shows that even such a high carbon price does not have a negative effect on the economy as a whole. Using econometric analysis, Metcaf (2015) also finds no statistically significant effect of the carbon tax on BC's economic growth. But what has been the impact on specific industries, as some sectors might have been more strongly affected than others? Rivers and Schaufele (2015b) show that BC's carbon tax has had no discernible impact, positive or negative, on the export of agricultural products. Instead, the carbon tax may well have been a factor in stimulating significant growth in the energy-related high technology sector in the province in recent years (Pedersen and Elgie 2015). In fact, between 2009 and 2011, revenue generated by the clean technology sector in BC grew by 17 percent, and the sector remains a locus of significant expansion. Yamazaki (2017) explores the employment impacts of the BC carbon tax and indicates that in clean service industries employment has increased while it has decreased in most carbon-intensive and trade-sensitive industries. In aggregate across industries, he finds the BC carbon tax to generate, on average, a small but statistically significant 0.74 percent annual increase in employment over the period 2007–13.

For households, the data shows compellingly positive effects. Lee (2011) looks at the distributional effects of BC's carbon tax and the revenue recycling to households in each income decile for 2010 and 2012. He shows that the income top 10 percent, on average, receive net benefits of up to 1 percent in 2012, while the bottom 10 percent face net costs of 0.5 percent. Still, the total effect on households is slightly positive. Melton and Peters (2013) expand on the previous analysis by providing a comprehensive assessment of impacts on households. In particular, they account for how the carbon tax and its revenue recycling regime affects households by altering investment and economic activity across the province. As a result, they found that benefits of personal and corporate tax cuts outweigh the costs of the carbon tax. In particular, due to lower corporate taxes, the province is likely to become more economically competitive relative to other jurisdictions in North America. The improved competitiveness has a positive impact on the provincial economy and on household incomes in the long run.[1] Using a computable general equilibrium model, Beck et al. (2015) suggest that even prior to revenue recycling, the BC carbon tax is positive. The authors suggest that this finding is a result of the tax incidence falling partly on wages. Because low-income households derive most income from government transfers, they are insulated from falling real wages. Also, Beck et al. (2016) estimate the differential impacts of the tax on urban and rural households. They find that, without redistributive measures, rural households would be disadvantaged by the tax, but the introduction of the Northern and Rural Homeowner Tax Credit is sufficient to make these households net beneficiaries, on average.

In sum, considering the above-described design, the BC carbon tax fulfills to a considerable extent ambitious sustainability criteria:

- A broad tax base guarantees comprehensive coverage.
- The carbon price is higher than in most emissions trading schemes (World Bank/Ecofys 2016).
- Administration costs are minimized.
- Revenue neutrality is mandated and ensures full, transparent, ongoing revenue recycling to BC households and businesses.
- Low-income households and disadvantaged communities are especially protected.

As a consequence, empirical data and simulation models suggest that BC's carbon tax combined with the specific revenue recycling scheme has reduced per capita GHG emissions by 10 percent without any detrimental effect on the economy or income distribution.[2]

## 2.2 The Politics

The BC case shows how a 'perfect storm of political conditions' (Harrison 2013a) can lead to the successful implementation of meaningful carbon pricing.[3] In BC, though, individual actors' behavior was heavily influenced by a number of external factors. First, BC's energy and GHG emission structure allows for high carbon prices and favors a tax solution (BC Gov 2012). BC commands a large potential of hydro power, which provides 93 percent of BC's electricity and leads to Canada's second lowest carbon intensity of 12.4 kg $CO_2e$ emissions per capita per year. In addition, stationary sources only account for 37 percent of BC's total GHG emissions, while transport accounts for 35 percent. BC has economically significant GHG-intensive industries such as cement and aluminum, but the local coal industry, which accounts for 80 percent of Canada's coal production, exports almost all of its coal to other jurisdictions. This particular situation translates into fewer vested interests in fossil fuel use, which in turn reduces political opposition from heavy polluters. Also, a tax solution appears particularly palatable given the importance of transport emissions.

Second, several US states and even the US federal government, an important point of reference for Canada, were working on state-based cap-and-trade schemes. Led by California, the Western Climate Initiative (WCI) formulated a model rule for a comprehensive linked cap-and-trade program; an initiative that was also joined by BC. And the US Northeast implemented the Regional Greenhouse Gas Initiative (RGGI). This provided hope that others would follow the BC example.

Third, several major public events were raising public attention to the issue of global warming in the 2000s (Rudolph 2005). Hurricane Katrina devastated the US Gulf Coast, the Stern Review showed the benefits of immediate climate action, Al Gore presented 'An Inconvenient Truth', the 4th Intergovernmental Panel on Climate Change (IPCC) received widespread public attention, and Gore and the IPCC even received the 2007 Nobel Peace Prize.

More importantly, fourth, by 2012 a beetle infestation had destroyed 50 percent of the stock of commercially valuable BC pine, causing severe damage in the billions to the BC economy (Pedersen and Elgie 2015). Temperature increases were identified as the major cause. In addition, due to the red-brownish coloring of the dead trees e.g. in Vancouver's treasured Stanley Park, the damage was easily visible to the BC public, for whom BC's nature and particularly its vast forests are a cherished resource. This made climate change a decisive political issues at the relevant time.

Fifth, the specific political system in BC favored a strong premier and allowed vote maximization on environmental issues by a liberal, right-of-center party (Harrison 2013a). The combination of a parliamentary government and a single-member plurality electoral system concentrates political power in a small number of hands, usually the leader of the party that holds the majority of seats. In addition, not being threatened by vote losses to extreme right parties, the right-of-center Liberal Party (LP) could hope for winning central-left votes by engaging in environmental policy, usually an issue more strongly associated with left-of-center parties. This made the political push for an ambitious carbon tax led by the premier a promising strategy.

Partly as a reaction to these external conditions, partly out of self-interest, political stakeholders in BC behaved in a way that cleared the way for the introduction of the BC carbon tax (Jaccard 2012). In the early 2000s, BC's voting public showed support for an ambitious GHG pricing (Harrison 2013a). In general, a majority of British Columbians see environmental protection as a hallmark of their province and BC as the environmental leader in Canada. At the time in question, above all, they were alarmed about the destruction of BC's forests. The economic loss caused by the beetle plague, but especially the immediate perceptibility of the damage to the BC pine trees, created significant public pressure on politicians to act on climate change.

However, despite British Columbians' generally high environmental awareness and their dissatisfaction with preceding climate policy in Canada and their own province, it was only the early 2000s that saw environmental issues becoming the most important problem, surpassing health care and the economy as leading concerns.

As a consequence, despite a general skepticism towards new taxes, the carbon tax proposed in February 2008 found the support of a majority of British Columbians in early opinion polls. However, the phase between the announcement of the tax and the tax taking effect in July of 2008 saw a decline in support. This decline was mainly caused by a gasoline price peak of more than 1.50 C$ in Vancouver, which was in fact caused by an increase in the international crude oil price, and additionally fired by special interests claiming the carbon tax to be unfair, amongst them advocates for the poor, northern residents who live in a colder climate, and rural residents with no access to public transportation. Although all these claims were proven wrong by scientific studies and a majority of BC households were in fact overcompensated by income tax cuts and direct payments to low-income households, support rates remained low, a fact that is attributed by some observers to the lack of trust in government, the irrational refusal of taxes, and a lack of understanding of the revenue recycling process. Luckily, the discontent was neutralized by voters' attention being drawn to the threat of a global recession after the collapse of the housing bubble in the US in 2008.

Politicians were aware of public concern for their immediate environment and their call for climate action (Harrison 2013a). Due to the specific political system in BC, the ruling right-of-center Liberals could not lose votes to the political right, so fishing for votes on the left side of the political spectrum, e.g. centrists votes of the New Democratic Party (NDP), by engaging in climate protection, represented a promising strategy. But anyway, after a resounding victory in the 2001 election, a reconfirming second victory in 2005, and the next election only being held in 2009, the Liberals commanded enough political leeway to drive the climate policy agenda. Even the early 2008 'Axe the Tax' campaign of the opposing NDP could not hurt the Liberals persistently, because voters put more trust in the Liberals to cope with the emerging global economic crisis from late 2008 onwards. In addition, the Liberals did not have to fear too much resistance from the business community as, first, companies traditionally had trusted liberal, right-of-center parties more than left-of-center parties, and, second, there was no alternative to the right of the Liberals for businesses to pin their hopes on.

In particular, BC Premier Gordon Campbell was strongly committed to act against climate change. Although formerly not known for his environmental agenda, but rather for his focus on economic growth and the government's budget, between 2006 and 2008, after five years in office, Campbell decided to forcefully act on global warming and establish himself as the leader in the Canadian debate on market-based climate protection. He established important climate policy institutions such as the Cabinet Committee on Climate Action and the Climate Action Secretariat, and he vigorously pushed for the carbon tax. In 2008, during the critical phase between the carbon tax announcement in February and the implementation in July, Campbell even put his own career at stake by tying his own political fate to that of the carbon tax.

The left-of-center NDP, on the other hand, gambled for high stakes as there was a serious threat of losing environmentally minded voters to the LP at the dawn of the 2009 elections. So, even after a resolution in favor of a carbon tax in early 2008, capitalizing on the interim drop in popularity of the BC carbon tax in late spring 2008, the NDP started a year-long campaign against the tax. The martially termed 'Axe the Tax' campaign argued – against the obvious facts – that the BC 'gas tax' was unfair as it would burden working people, while industry would get away with paying nothing. Instead, the NDP favored a cap-and-trade scheme focused on big polluters and promised to repeal the tax if voted into office. The campaign, though met with disbelief by the environmental community, reached its goal with decreasing support for the carbon tax amongst British Columbians and a surge forward in opinion polls for the NDP in summer and fall of 2008. Opinion polls even show that the carbon tax was the decisive issue for that.

British Columbia traditionally has a strong environmental movement, with Greenpeace being found in Vancouver and other strong players like the David Suzuki Foundation having won important victories over economic interests. Environmental organizations favored ambitious climate action and a market-based approach. The 2007 climate package of the BC government was largely supported by BC environmental groups. However, they called for even more ambitious targets and the use of market-based approaches. For the 2008 budget, the environmental community demanded a comprehensive carbon tax as the major means for reaching BC's GHG reduction target. Their claim was even supported by a group of 70 economists, who urged the BC government to introduce a carbon tax. Once the government proposal for a carbon tax covering almost 80 percent of BC's GHG and increasing to a level of 30 C$ per ton in only four years was on the table, the environmental community almost unanimously hailed the approach.

Industry, at least at the beginning, was not actively opposed to the BC carbon tax (Harrison 2013a). After the announcement of the 2007 BC climate program, the business community was afraid of being confronted with severe measures and high costs in order to achieve the 33 percent reduction, while it already expected some kind of carbon pricing, especially considering similar developments all across North America. Businesses had, however, confidence in the liberal government not to put too high a burden on companies. Actually, they had sent clear signals that a BC carbon tax was to equally take into consideration transport, household, and industry emissions and had to be revenue neutral, a wish that was granted by the specific design of the tax. The reduction of the corporate tax, comprehensive coverage, a tax rate escalator with a low start, predictable tax hikes, and a price limit at the end were design features that particularly appealed to the business community and hence reduced their taste for opposition. Still, after the official announcement of the tax in February 2008, some special interest groups raised concerns, particularly trucking firms, farmers, and energy-intensive industries. However, again these concerns were rejected by the government, based on scientific evidence.

The public administration, in particular the Climate Action Secretariat and the Ministry of Finance (MoF), were strong supporters of the carbon tax idea. While the Secretariat mainly saw the environmental merits, MoF also hoped to have a say in climate policy due to the tax character of the policy and the need for a revenue spending strategy. Both bodies expected carbon pricing to be spreading across North America anyway, so BC would not be alone for long and might even benefit from having been the leader in the discussion. Particularly, the MoF favored a tax as it made a broad coverage easier, was quick to implement, and easy to administer using the existing institutional structures. Both institutions' influence in the design was particularly strong, as the carbon tax was presented as an accomplished fact to interest groups and the general public, without any leeway for public debates or rent seeking.

In sum, the confluence of reinforcing external factors and influential political actors not being as opposed as expected – with voters, ruling politicians, and the environmental administration even being outright supportive – enabled BC to implement a carbon tax close to 'the economist's ideal' (Jaccard 2012: 182).

Despite its success and strong hopes of being followed by other provinces, for five years, from 2008 to 2012, BC remained the only Canadian province to apply explicit and ambitious carbon pricing. However, 2013 saw the beginning of new dynamics in sub-national carbon pricing in Canada, culminating even in a new federal-level carbon pricing initiative.

# 3 FOLLOWING SUIT: QUÉBEC'S AND ONTARIO'S CARBON MARKETS AND THE TRUDEAU INITIATIVE

## 3.1 Carbon Market Design and Evaluation

Québec established its carbon market in 2013, while Ontario's program only came into operation in 2017 with a linking option in 2018. Both the Québec and the Ontario cap-and-trade system basically operate under the umbrella of the WCI, and hence share many design features (Gouvernement du Québec 2014; Ontario Ministry of the Environment and Climate Change 2017). Participation in both programs is mandatory for all targeted entities with emissions of at least 25,000 t $CO_2e$ per year. Both carbon markets cover all Kyoto GHG plus $NF_3$. Stationary sources like large industrial facilities, power generation, and electricity imports are covered at the emitting facility level, while transport fuels and natural gas for the transport sector and households are – in Québec after a major program expansion in 2015 – covered at the distributor level. Hence, 80 entities in Québec and 150 in Ontario are subject to cap-and-trade rules and about 82–5 percent of the two provinces' GHG emissions are capped. Québec's cap was 65 m t in 2015 and it will be reduced by 1–2 percent annually to a final level of 55 m t in 2020, which is about 25 percent below 1990 emissions of the covered sectors. In Ontario the 2017 cap was set at 142 m t and will be reduced by around 4 percent in the following two years to a final level of 125 m t, a level about 15 percent below 1990 emissions.

Québec and Ontario hand out allowances each worth one metric ton of $CO_2e$ emissions in a given year. One hundred percent of allowances needed for compliance have to be obtained at auctions or the secondary market by non-industry sectors such as electricity generators and importers, natural gas and transport fuel distributors. However, industries with combustion-based emissions and subject to international competition receive up to 100 percent of their allowances for free.[4] Since 2014 Québec has held non-discriminatory auctions jointly with California in February, May, August and November, while Ontario still holds its own quarterly auctions. The market is open to all covered facilities and interested parties. Allowances can also be obtained on secondary markets such as the Intercontinental Exchange.

In Québec and Ontario all auction revenues go to funds specifically dedicated to the reduction of GHG emissions, the Québec Green Fund and the Ontario Greenhouse Gas Reduction Account. While there is no legal obligation, there is also on option for using revenues for mitigating negative social impacts of emissions reduction efforts.

In both Québec and Ontario banking is allowed without limits, while borrowing is prohibited. Offsets are allowed for coverage of up to 8 percent of individual entities' compliance obligations. Reductions have to be additional, durable and verifiable, and offsets will be verified by independent agents. In Québec, only province-based projects of ozone-depleting substance destruction, manure storage methane destruction and landfill site gas capture are eligible, while Ontario only accepts credits from projects outside of the province.

For cost containment Québec and Ontario have established a price collar that is aligned with the rules of the bigger WCI partner, California. Acting as a price ceiling, allowances are sold in three equal price tiers at 40, 45 and 50 C$ (2013).[5] Representing the price floor, a minimum auction price of 10 C$ (2013) was set. Both upper and lower price limits have increased by 5 percent above inflation every year since 2013.

Québec's compliance periods are 2013–14, 2015–17 and 2018–20, while Ontario's first and so far only compliance period lasts from 2017 to 2020. However, facilities have to report their GHG emissions annually and obtain independent emission report verification. Emissions as well as allowances are registered in tracking systems. In the case of non-compliance, facilities face fines of up to several million C$ per case, imprisonment, and suspension from any new allowance allocations. In addition, three allowances have to be surrendered for each non-covered ton of emissions.

A link has already been established between Québec and California under the WCI umbrella, enabling the mutual acceptance of emission allowances for compliance. Ontario, also a WCI member from the beginning, intends to join in 2018.

Evaluating the design of the Ontario and Québec cap-and-trade programs against ambitious sustainability criteria provides an ambivalent picture. On the positive side of both programs,

- coverage is comprehensive;
- flexibility mechanisms feature a sustainable design;
- the compliance system is reliable.

However, on the negative side,

- caps lack ambition;
- bigger parts of industry are exempted from auctioning;
- revenue use does not reflect regressive carbon price effects on low-income households;
- while a price collar applies, both price floor and price ceiling are too low.

While it is rather early for an *ex post* analysis of the effects of the Québec and Ontario cap-and-trade programs, especially in the Ontario case, two major results have already become obvious – and are also supported by the results from the WCI partner, California. Compliance is close to 100 percent, which means that emissions stay below the cap. Allowance prices just slightly exceed the reserve price, which underlines that there is a lack of scarcity in the market and significant leeway for cap cutbacks.

## 3.2 The Politics

As outlined above, BC's leadership in ambitious carbon pricing in Canada became possible due to a unique confluence of external factors and political stakeholder behavior. However, until 2013, despite the BC success, no other Canadian province followed suit nor did the federal government engage in carbon pricing.

Harrison (2013b), mainly referring to US examples such as Californian tailpipe or clean energy standards, while also mentioning some Canadian cases, including the participation of Canadian provinces in the WCI, proposes a set of reasons for the limits to leadership in sub-national carbon pricing: First, sub-national jurisdictions focus on those emitters and emissions that promise immediate local benefits. And due to the differences in the economic and emissions structure between sub-national jurisdictions, simply following the good examples of the leaders appears unwise. In Canadian practice for example, the provinces with the lowest carbon intensity such as Québec or BC have been the most active in ambitious carbon pricing.

Second, product standards are preferred to process standards, which might shift burdens to other jurisdictions. In Canada, as an example, California tailpipe standards for motor vehicles' GHG emissions were supported by nine provinces in 2007, the only exemption being Ontario, a province relying heavily on car manufacturing.

Third, sub-national jurisdictions try to claim credit for emission reductions in other jurisdictions. California, for example, applies its clean electricity standard to out-of-state producers, but accounts for respective emission reductions in its emissions inventory and allows Californian utilities to use these credits for meeting their cap-and-trade obligations.

And fourth, sub-national jurisdictions minimize negative impacts by inter-jurisdictional collaboration or lobbying for federal intervention; a lack of either, at least in the long run, hinders sub-national activities. In North America, while New York State, California, and BC leadership had been crucial in the discussion on national carbon pricing schemes, WCI and RGGI members lobbied at the federal level for a nationwide carbon pricing scheme, which eventually failed in both the US and Canada in 2010, leading to the breakup of the biggest part of the WCI coalition.

However, Harrison (2013b) also appreciates three major streams of supportive inter-jurisdictional dynamics. First, there appears to be a 'classical pattern' of policy idea diffusion, apparent in North America, for example, in Renewable Portfolio Standards, tailpipe standards for motor vehicles' GHG emissions, or low-carbon fuel standards. Second, political goodwill to act as leaders seems to play a decisive role in sub-national climate policy guidance, obvious in the US case of California and the Canadian case of BC. Third, inter-jurisdictional coordination reduces the risk of leakage and competitive disadvantages and may pave the way for federal action, with RGGI and the WCI being good examples.

In the light of these arguments, current developments in Canada appear promising. Considering the limits, though still the province with the third lowest per capita GHG emissions but the second biggest emitter of total GHG emissions amongst Canadian provinces, Ontario joined WCI efforts to establish a comprehensive carbon market. Also, with Ontario applying carbon pricing, product standards appear to lose their dominating role. Not least, inter-jurisdictional collaboration appears to gain momentum with the revitalization of the WCI and intentions of even additional Canadian provinces and US states such as Manitoba and Washington State to re-join and link planned domestic carbon markets with WCI partners (Rudolph et al. 2017). In addition, some of the promising dynamics gain momentum. The idea of carbon cap-and-trade is spreading across all governance levels, including the regional and local level (ICAP 2018), and current carbon prices in the four Canadian provinces with explicit pricing schemes already range from 30 C$ in BC and 17 C$ in Ontario to around 15 C$ in Alberta and Québec. Political good will is showing on the subnational level in North America despite, or maybe because of, the Trump administration's withdrawal from ambitious climate policy. But most importantly for Canada, Trudeau's carbon pricing initiative provides the necessary federal-level incentive to engage in ambitious carbon pricing at the provincial level.

On October 3, 2016 Prime Minister Justin Trudeau enlivened the debate around carbon pricing by setting a deadline and a benchmark for province-level price levels. He warned the provinces in a speech on the Paris Agreement in the House of Commons that '[i]f neither price nor cap-and-trade is in place by 2018, the government of Canada will implement a price' (CBC 2016). Provinces' carbon prices should start at a minimum of 10 C$ in 2018 and increase by 10 C$ per year to a level of 50 C$ in 2022. Trudeau left it up to the provinces whether they want to choose taxes or a cap-and-trade program to implement this price. Guaranteeing 100 percent revenue neutrality for the federal government, however, he promised that all proceeds from provincial taxes or allowance auctioning would stay in the respective provinces. By December 2016, eight of the ten provinces had agreed to the plan, while mining-strong Saskatchewan remained the major objector. On May 18, 2017 the Canadian Environment Minister then published a concrete federal backstop plan, which is a mix of a carbon tax for consumers and a cap-and-trade system for major industry.

The current Canadian government considers carbon pricing a major means of achieving the national 30 percent GHG reduction below 2005 levels by 2030. The Trudeau Initiative is certainly a major step forward towards a nationwide carbon price in Canada. And while the 30 percent overall reduction target does not appear sufficiently ambitious, the pricing target of 50 C$ by 2022 is challenging. In addition, while federally mandating a certain minimum price level, giving provinces the freedom to choose the instrument allows them to act as policy laboratories.

## 4 CONCLUSIONS

The Paris Agreement is a major step towards limiting global warming to an acceptable level, but needs to be further strengthened in order to achieve the 2°C target. Carbon pricing is still the preferred policy strategy for reaching ambitious climate targets at lowest cost to society. However, there are still significant political barriers to sustainable carbon pricing schemes.

Canada is one of the biggest emitters of total and per capita GHG in the world. However, several provinces have successfully implemented carbon pricing schemes. BC has by far the most sustainable program, followed by Québec and Ontario. The 2016 Trudeau Initiative is going to force a carbon price from 10 C$ in 2018 to 50 C$ by 2022, which is an important step towards sustainable carbon pricing in Canada. Politically, BC leads the way due to a confluence of supportive external factors and stakeholder activities, which also show the limits of Public Choice analyses. Due to significant differences in the economic structure and the political power balance between provinces, combined with changing general circumstances, other provinces did not follow suit before 2013; and neither did the federal Canadian government. However, the late 2010s are seeing new dynamics in several provinces and even at the federal level.

Hence, the Canadian example shows how tailor-made sub-national action can supplement national action, especially in countries with reluctant federal governments. Under certain circumstances, sub-national action can stimulate not only neighboring jurisdictions, but also federal governments to follow suit. With federal US climate policy being paralyzed by the Trump administration, Canada, in collaboration with progressive states in the US, now has a historic chance of becoming a global leader in market-based climate policy development from the bottom up.

# NOTES

* Supported by The Japan Society for the Promotion of Sciences Grants-in-Aid for Scientific Research (KAKENHI No. KK20160009).
1. However, it is necessary to note that this results focuses on the impact of the carbon tax on the average BC household.
2. However, it is worth noting that the recent redistribution of some tax revenues to particular industries instead of broad-based tax cuts may reduce the overall cost-effectiveness and fairness (Murray and Rivers 2015).
3. Common explanations of sub-national climate policy leadership emphasize immediate environmental co-benefits, local economic benefits, and voters valuing climate policy leadership despite marginal environmental effects (Engel 2006).
4. However, this share is supposed to decrease over time and be thoroughly reviewed for the period after 2020.
5. But only entities registered in the respective provinces are eligible.

# REFERENCES

BC Gov (2012): British Columbia Greenhouse Gas Inventory Report 2010. Victoria.

Beck M, Rivers N, Wigle R, and Yonezawa H (2015): Carbon Tax and Revenue Recycling: Impacts on Households in British Columbia. *Resource and Energy Economics* 41: 40–69.

Beck M, Rivers N, and Yonezawa H (2016): A Rural Myth? Sources and Implications of the Perceived Unfairness of Carbon Taxes in Rural Communities. *Ecological Economics* 124: 124–34.

CBC (2016): Justin Trudeau gives provinces until 2018 to adopt carbon price plan. http://www.cbc.ca/news/politics/canada-trudeau-climate-change-1.3788825.

Elgie S and Mcclay J (2013): BC's Carbon Tax Shift Is Working Well after Four Years. *Canadian Public Policy* 39: 1–10.

Endres A (2011): Environmental Economics. Cambridge: Cambridge University Press.

Engel K (2006): State and Local Climate Change Initiatives. *Urban Lawyer* 38(4): 1015–29.

Gouvernement du Québec (2014): Québec's Cap-and-Trade System for Greenhouse Gas Emission Allowances. Québec.

Gulati S and Gholami Z (2015): Estimating the Impact of Carbon Tax on Natural Gas Demand in British Columbia. Sustainable Prosperity.

Harrison K (2013a): The Political Economy of British Columbia's Carbon Tax. OECD Environment Working Papers No. 63. Paris: OECD Publishing.

Harrison K (2013b): Federalism and Climate Policy Innovation. *Canadian Public Policy* XXXIX, Supplement 2, 95–108.
ICAP (2018): Emissions Trading Worldwide: Status Report 2018. Berlin: ICAP.
Jaccard M (2012): The Political Acceptability of Carbon Taxes. In: Milne J and Andersen MS (eds): Handbook of Research on Environmental Taxation. Cheltenham, UK/Northampton, USA: Edward Elgar Publishing, 175–91.
Lee M (2011): Fair and Effective Carbon Pricing Lessons from BC. Vancouver: Canadian Centre for Policy Alternatives.
Melton N and Peters J (2013): Is British Columbia's Carbon Tax Good for Household Income? Vancouver: Navius Research Inc.
Metcaf GE (2015): A Conceptual Framework for Measuring the Effectiveness of Green Fiscal Reforms. GGKP Research Committee on Fiscal Instruments, Working Paper 07.
Murray B and Rivers N (2015): British Columbia's Revenue-neutral Carbon Tax. *Energy Policy* 86: 674–83.
Oates WE (2004): A Reconsideration of Environmental Federalism. In: Oates WE (ed.): Environmental Policy and Fiscal Federalism. Cheltenham, UK/ Northampton, USA: Edward Elgar Publishing, 1–32.
Ontario Ministry of the Environment and Climate Change (2017): Cap-and-trade Program Overview. https://www.ontario.ca/page/cap-and-trade-program-over view.
Ostrom E (2009): A Polycentric Approach for Coping with Climate Change. Washington DC: World Bank (Background Paper to the 2010 World Development Report).
Pedersen T and Elgie S (2015): A Template for the World: British Columbia's Carbon Tax Shift. In: Kreiser L, Skou Andersen M, Egelund Olsen B, Speck S, Milne J, Ashiabor H (eds): Carbon Pricing, Design, Experiences and Issues. Cheltenham, UK/Northampton, USA: Edward Elgar Publishing, 3–15.
Rivers N and Schaufele B (2015a): Salience of Carbon Taxes in the Gasoline Market. *Journal of Environmental Economics and Management* 74: 23–36.
Rivers N and Schaufele B (2015b): The Effect of Carbon Taxes on Agricultural Trade. *Canadian Journal of Agricultural Economics* 63: 235–57.
Rudolph S (2005): Handelbare Emissionslizenzen. Marburg: Metropolis.
Rudolph S, Lenz C, Lerch A, Volmert B (2012): Towards Sustainable Carbon Markets. In: Kreiser L, Yábar Sterling A, Herrera P, Milne JE, Ashiabor H (eds): Carbon Pricing, Growth and the Environment. Cheltenham, UK/ Northampton, USA: Edward Elgar Publishing, 167–83.
Rudolph S, Lerch A, Kawakatsu T (2014): Regional Market-based Climate Policy in North America. In: Kreiser L, Soocheol L, Kazuhiro U, Milne JE, Ashiabor H (eds): Environmental Taxation and Green Fiscal Reform for a Sustainable Future. Cheltenham, UK/Northampton, USA: Edward Elgar Publishing, 273–88.
Rudolph S, Lerch A, Kawakatsu T (2017): Developing the North American Carbon Market. In: Weishaar S, Kreiser L, Milne JE, Ashiabor H, Mehling M (eds): The Green Market Transition. Cheltenham, UK/Northampton, USA: Edward Elgar Publishing, 209–30.
Schneider F, Kollmann A, Reichl J (2015): Political Economy and Instruments of Environmental Politics. Cambridge: MIT Press.
Stewart RB (1977): Pyramids of Sacrifice? *The Yale Law Journal* 86(6): 1196–272.
UN (2015): Paris Agreement. New York.

UNFCCC (2016): Aggregate Effect of the Intended Nationally Determined Contributions (Synthesis report by the secretariat). Bonn.

World Bank/Ecofys (2016): Carbon Pricing Watch 2016. Washington D.C.

Yamazaki A (2017): Jobs and Climate Policy. *Journal of Environmental Economics and Management* 83: 197–216.

# 4. Beyond Thunderdome? The prospects of federal greenhouse gas cap-and-trade in Australia*

**Elena Aydos and Sven Rudolph**

## 1 INTRODUCTION

'All we want is life beyond the Thunderdome' is a line made famous by Tina Turner in the title song of the 1985 *Mad Max* movie. In Thunderdome, a gladiatorial arena in a post-apocalyptic world, conflicts are resolved by duels to death. Not as violently, but certainly as fiercely, political battles were fought in Australia over greenhouse gas (GHG) cap-and-trade in the early 2000s. Increasingly, the informed public and many of the political stakeholders have craved for a 'life beyond'.

With Australia having some of the highest per capita GHG emissions, being the world's leading coal exporter, and its political stakeholders still licking wounds from previous political battles, the question arises: Is there really a second chance for sustainable cap-and-trade in Australia, a 'life' beyond Thunderdome?

In this chapter, we answer this question by evaluating Australia's former cap-and-trade initiatives based on sustainability criteria. Against the background of Public Choice theory we then analyse the reasons for the political failure of earlier cap-and-trade in Australia and predict the chances of reviving any of the former approaches. We mainly argue that earlier cap-and-trade schemes in Australia showed merit and that, despite ongoing partisan feuding, there are political chances for cap-and-trade to live 'beyond Thunderdome'.

## 2 CLIMATE POLICY, SUSTAINABLE CAP-AND-TRADE, AND THE POLITICO-ECONOMIC BACKGROUND

The Paris Agreement is certainly a diplomatic success but it must be substantiated by convincing policies. The agreement explicitly allows the use of 'internationally transferred mitigation outcomes', or, in economic terms, the trading of emission rights or credits. Emissions trading or cap-and-trade has been almost unanimously supported by economists on the grounds of environmental effectiveness and economic efficiency.[1] Recently, it has been shown that a sustainable design, taking into account not only effectiveness and efficiency, but also social justice requirements, is possible.[2] Surprisingly, the resulting design recommendations do not exhibit major contradictions between environmental, economic and social goals, but rather point in the same direction. Following this approach would hence directly answer to the Paris Agreement's urge to 'reflect equity'.[3] Not least, linking domestic schemes can significantly improve the sustainability of cap-and-trade,[4] and with domestic schemes becoming more widespread, extending not only to several continents and countries, but also to all governance levels from local to supra-national,[5] this option becomes even more promising.

However, the implementation of sustainable cap-and-trade faces a variety of political barriers. Public Choice, the economic theory of politics, argues that rational and self-interested political actors such as voters, interest groups, bureaucrats and politicians would not support ambitious market-based approaches to environmental protection.[6] It concludes that there is a 'market tendency for the political process to resist market mechanisms for rationing scarce environmental resources'.[7] Empirical studies support this view to some extent.[8] Thus, while the restrictive assumptions of Public Choice certainly limit its explanatory and predictive power, Public Choice can still provide a worst-case scenario for the political feasibility of sustainable cap-and-trade.

## 3 CAP-AND-TRADE IN AUSTRALIA

In 2008, the Labor government under Prime Minister Rudd proposed the introduction of an emissions trading scheme (ETS) known as the Carbon Pollution Reduction Scheme (CPRS).[9] A thorough policy development process was put in place, starting with a Green Paper on ETS design issues in July. This was followed by the release of a comprehensive independent report on the impacts of climate change on the Australian economy,[10] the

Treasury modelling and a White Paper in December.[11] In 2009 and 2010, the Rudd Government introduced three packages of legislation to implement the scheme. The CPRS Bill passed the House of Representatives. In a historical political turn, six days before the Senate vote, Abbott won the opposition leadership replacing Turnbull and opposed the Bills.[12] The Green Party voted with Abbott against the CPRS.

After a second failed attempt in 2010, Prime Minister Rudd deferred the CPRS legislation until the end of the first commitment period of the Kyoto Protocol in 2012. The decision to postpone the CPRS Bills, the chief policy mechanism to deal with what Rudd called the 'moral challenge of our generation', cost Rudd his position as a Labor leader and Prime Minister.[13]

In July 2011, Prime Minister Gillard proposed the introduction of a carbon market called the Carbon Pricing Mechanism (CPM).[14] This time backed by the Green Party, the legislative package passed in Parliament in November 2011, and received Royal Assent in December 2011.[15] The CPM commenced on 1 July 2012, coinciding with the end of the first commitment period under the Kyoto Protocol. After two years of a functioning AUS CPM, Australia's new Prime Minister Abbott delivered a campaign promise of abolishing the scheme. A legislative package, also known as the 'carbon tax repeal legislation', passed in the House of Representatives and the Senate and received Royal Assent on 17 July 2014, dismantling the AUS CPM. The legislation entered into effect from 1 July 2014.

Following the abolition of the CPM,[16] the Coalition Government's main climate change policy instrument became the Emissions Reduction Fund (ERF), which was built upon the already existing legal framework for the Carbon Farming Initiative (CFI) through amendments to the CFI Act.[17] The ERF is an incentive-based scheme, in which the federal government subsidises sequestration or emission avoidance projects through the direct purchase of offset credit units known as Australian Carbon Credit Units (ACCUs).[18] The process for purchase of ACCUs is completed via reverse auctions and tenders,[19] whereby project proponents compete for the undertaking of emissions abatement projects. The purchaser (in this case, the government), selects the successful bids to enter into a contract with.[20]

Linked to the ERF is the Safeguard Mechanism (SM), which commenced operations on 1 July 2016 with the passing of the National Greenhouse and Energy Reporting Act 2007 (Cth) and respective amendments made under the CFI Amendment Act.[21] The SM is a baseline and credit mechanism covering a relatively small number of high emitters in Australia. The key features of the CPRS, CPM and ERF/SM schemes are compared below and assessed against the sustainability criteria for cap-and-trade.

### 3.1 Coverage

The sustainability criteria on coverage call for mandatory participation of all polluters and coverage of all GHG.[22] Such requirements are only partially fulfilled by the CPRS and the CPM, with the CPRS complying with the criteria to the greatest extent, while the ERF/SM fails to comply with the social justice criteria in relation to coverage.

The CPRS fulfilled the mandatory participation criterion and GHG coverage criterion, covering all six GHG listed under the Kyoto Protocol, while only partially meeting the requirements in respect to participation of all polluters. The CPRS was designed to be mandatory for approximately 1,000 large polluters from the stationary energy, transport, fugitive emissions, industrial processes and waste sectors, covering approximately 75 per cent of Australia's emissions.[23] Within these sectors, only companies emitting 25,000 tonnes or more of $CO_2$-e per year were covered.[24] Excluded from the CPRS were emissions from agriculture, forestry, fugitive emissions from decommissioned underground coal mines, certain synthetic GHGs and emissions from the combustion of biomass.

The CPM was also mandatory for liable companies. It covered approximately 360 large polluters emitting 25,000 tonnes of $CO_2$-e per year or more, responsible for around 60 per cent of Australia's emissions.[25] The CPM also covered only four of the six Kyoto GHG, from aluminium smelting, stationary energy, non-legacy waste, transport,[26] industrial processes and fugitive emissions. The scheme excluded emissions from agriculture, forestry, fugitive emissions from decommissioned coal mines and legacy waste. Road transport and forestry sectors were not covered by the CPM. However, the legislation on fuel tax and synthetic GHGs imposed an equivalent carbon price on some business transport emissions, the non-transport use of liquid and gaseous fuels (except natural gas) and synthetic GHGs.[27]

Of the three schemes, the ERF/SM is the one with the most significant sustainability problems. The voluntary participation in the ERF coexists with a mandatory participation for a very limited number of businesses in the SM, approximately 140 large businesses with annual emissions of over 100,000 tonnes of carbon dioxide equivalent (compared to 25,000 tonnes of $CO_2$-e per year or more under the CPRS and CPM). The ERF/SM covers exclusively direct GHG emissions (scope 1 emissions) from power generation (approximately 57 per cent GHG emissions from the electricity sector), mining (coal and metal ores), oil and gas extraction, gas supply, manufacturing (including metals, cement and lime), transport (air, sea, rail and road), heavy and civil engineering construction, and (new) waste.

## 3.2 Cap

Sustainable cap-and-trade have stringent volume caps, which are absolute and are set to gradually decrease and are linked to stringent absolute volume reduction targets.[28] The criteria are partially fulfilled by the CPRS and the CPM, with the ERF/SM lagging behind.

Australia has historically committed to weak reduction targets and the trend has been further exacerbated under the recent Coalition Government. At the time of the CPRS proposal, Australia had committed not to a reduction, but to an increase in emissions to average 108 per cent of 1990 emissions for the years 2008 to 2012. Beyond 2012, reduction targets were set at 60 per cent below 2000 levels by 2050 and between 5 per cent and 15 per cent below 2000 levels by 2020. In the event that there is comprehensive international agreement, the ambition of the target could be increased to 25 per cent below 2000 levels by 2020. In terms of cap, while transitional measures would exclude an absolute cap in the first 12 months of the CPRS, from 2012–13 absolute volume caps would be set by regulations, based on the indicative national targets in the relevant year.

Functioning under the same federal emission reduction targets, the framework for the CPM provided for a longer transitional period (three years) during which emissions would not be capped under the scheme, disproportionately burdening non-covered sectors.[29] However, from 1 July 2015 onwards, absolute pollution caps would be set by regulations[30] reflecting Australia's medium- and long-term targets for reducing net GHG emissions and the different possible trajectories towards it.[31]

With the absolute caps being set by regulations, there was no mandatory cap reduction under the CPRS and the CPM. However, absolute caps were expected to gradually decrease, amongst other things, through the adoption of default caps in the absence of regulations.

The Coalition Government has committed to the lowest level of reduction targets for 2020 (only 5 per cent reduction based on 2000 levels) and adopted a weak medium-term target of 26–8 per cent below 2005 levels by 2030. Under the ERF/SM there are no absolute caps. Instead, individual baseline emissions numbers are set for each facility, calculated based on historical emissions data (highest level of reported emissions for a facility over the historical period 2009–10 to 2013–14), with no mandatory graduation reduction of individual baselines. The ERF/SM once again violates the social justice criteria in relation to the cap.

### 3.3 Allocation

The social justice criteria for the validity of emission rights and the initial allocation would be increasingly fulfilled by the CPRS and the CPM. Both schemes would create and issue emissions permits, each corresponding to 1 tonne of $CO_2$e. During the first 12 months of the CPRS (2011–12), permits would have been allocated at a fixed charge of $10 per unit. From 2012–13 there would be auctioning of permits (unit of 1 t of $CO_2$e/a) combined with targeted assistance via free allocation of permits to emissions-intensive trade-exposed (EITE) industries and the coal sector. Assistance would be transitional and the scheme would progressively move towards 100 per cent auctioning.

The framework of the CPM provided for the issue of permits for a fixed price from 1 July 2012 until 30 June 2015, starting at $23 per tonne of $CO_2$-e in 2012–13. After 1 July 2015, units would be allocated via auction. During the fixed charge years and the flexible charge years, the government would issue permits free of charge to EITE sectors under the Jobs and Competitiveness Program. Two categories of eligibility for the Jobs and Competitiveness Program, i.e. moderately emissions-intensive and highly emissions-intensive, would determine the different levels of free allocation.[32]

In contrast, the sustainability criteria for allocation are entirely violated by the ERF/SM. Under the ERF/SM there is no absolute cap and no issue and/or trading of permits. The federal government enters into 'carbon abatement contracts' to directly purchase offset credits generated through sequestration or emission avoidance projects.[33]

### 3.4 Revenue Use

The CPRS and the CPM largely complied with the sustainability criteria for revenue use, as both schemes were set to be revenue neutral. Revenue from the CPRS would have been used to purchase international credits, such as avoided deforestation credits, and towards household assistance measures and industry assistance measures.

The CPM provided for cost compensation to low-income households and industry, investments in renewable energy and funding for emissions reduction projects in the land sector. Over 50 per cent of carbon price revenue was earmarked for cost compensation of approximately 1 million low-income households. The package effectively delivered a tax reform that compensated beyond the cost increase due to the carbon price.[34] In addition to tax cuts, pensions, allowances and benefits were increased and there were other benefits to households with special needs.

The Jobs and Competitiveness Program, providing free allocation of permits to EITE sectors, would use another 40 per cent of the revenue collected under the CPM. A Coal Sector Jobs Package and an Energy Security Fund would guarantee free allocation of permits and cash payments to the coal sector, including coal mining and coal-fired electricity generators. A Steel Transformation Plan provided assistance to the Steel sector. Finally, part of the revenue from the CPM would be invested in renewable energy, low pollution and energy efficiency technologies, as well as funding new land-based mitigation measures.

Contrarily, the ERF/SM is a subsidy-based scheme that uses government revenue to purchase offset credits and therefore it is not capable of generating revenue.

## 3.5 Flexibility Mechanisms

The sustainability criteria in respect to flexibility mechanisms allow for banking but disapprove of borrowing of permits.[35] A limited number of offsets that meet stringent requirements is also in line with the sustainability criteria.

The CPRS and the CPM complied to some extent with the criteria, allowing for banking of units after the fixed charge period, but also allowing for limited borrowing. Under the CPRS, a domestic offsets program offered the opportunity to receive free Australian emissions units for sustainable offset projects. The CPRS would also accept international offset units, including certified emission reduction (CERs), emission reduction units (ERUs), removal units (RMUs), prescribed Kyoto units and prescribed non-Kyoto international emissions units.[36]

The CPM was linked to the CFI, a domestic voluntary offsets scheme offering a range of abatement and carbon sequestration opportunities in the land sector. The CPM would also link to international schemes from 2015 onwards, up to a limit of 50 per cent of the participants' liability for the relevant year.[37]

Under the ERF/SM, facility operators can surrender eligible carbon offsets at any time to remain below their baseline. Credits issued under the Emissions Reduction Fund – also known as Australian Carbon Credit Units or ACCUs – are eligible offsets.

## 3.6 Price Management

The sustainability criteria suggest that market intervention should be kept to the minimum, but if a price collar applies, the price floor and the price ceiling should not be lower than US$20 and US$200, respectively.[38] The

CPRS and the CPM would progressively fulfil this requirement, as price flexibility was meant to increase over time.

In the first 12 months of the CPRS, permits would have been sold for a fixed price, with a practical effect of a carbon tax. From 2012–13, permits would be auctioned. In the first four years of auctioning, access to an unlimited store of additional permits issued at a pre-specified fixed price would have the practical effect of a price cap, starting at AU$40. These units would not be tradeable or bankable for future use.

Similarly, under the CPM permits were sold during the fixed-charge years at AU$23 per permit. A price ceiling was in place for the first three flexible charge years.[39] The original design of the AUS CPM also included a price floor (AU$15, rising annually by 4 per cent) in the first three flexible charge years.[40] However, this feature was removed when Australia and the European Union declared that they would link the CPM with the European Union Emissions Trading System (EU ETS).[41]

It is not possible to assess the ERF/SM against this criterion, given that permits are not issued by the government under this scheme.

## 3.7 Compliance

According to the sustainability criteria, compliance periods should be short in order to allow for short-term control over reduction achievements and provide opportunities for immediate penalties and *ex post* emission compensation in the case of non-compliance. Trading periods can be long only if supplemented by short-term submission requirement for major parts of used emissions rights. Reliable monitoring and quenching penalties are a necessary component of a sustainable cap-and-trade.[42] The CPRS and the CPM have to a large extent complied with this criterion, while, once again, the ERF/SM fails to comply.

The compliance cycles of the CPRS and the CPM were the financial year. Both schemes imposed stringent penalties in case of unit shortfall, but no (over)compensation for excess emissions. Penalties were strict under the CPM, with a unit shortfall charge applicable in the first three years (fixed charge period) equivalent to 130 per cent of the fixed price for Australian Carbon Units (ACU), once again not including (over) compensation for excess emissions.[43] In flexible charge years, the unit shortfall charge was set by regulations and would range between 130 per cent and 200 per cent (default rate) of the benchmark average auction charge for the relevant period.[44] In case a unit shortfall charge remained unpaid after the due date, an extra penalty calculated at the rate of 20 per cent per annum (or a lower rate specified in the regulations) on the amount unpaid was due.[45]

In addition to the generous baselines and very limited liability under the SM, the ERF/SM has very weak compliance mechanisms. Participants may opt between one financial year or multi-year periods (two- or three-year multi-year periods). The penalty for exceeding the baseline is virtually insignificant, with the option for the participant to adjust the facility's baseline or select a multi-year compliance approach for managing excess emissions.

The Regulator has discretion in applying a range of enforcement options, including the issuing of infringement notices, acceptance of enforceable undertakings, seeking injunctions and pursuing court action. Enforcement options are unlikely to ever be applied given the generous baselines. The Clean Energy Regulator may seek civil penalties through the courts with the maximum amount set at 100 penalty points per day (currently $18,000 per day), to a maximum of 10,000 penalty points in total. In addition to paying the penalty, the facility operator remains under an obligation to rectify an excess emissions situation. This option is also very unlikely to take place, as the civil penalty is considered a last resort and will never apply to businesses that meet legislated safeguard requirements.

## 3.8 Supporting Measures

None of the Australian schemes to date provided for border adjustment to prevent carbon leakage. The main strategy to protect EITE sectors in the CPRS and the CPM was free allocation of permits, which does not comply with the sustainability criteria.

In August 2012, the linking of the EU ETS and Australia's CPM was announced. GHG emissions permits from the EU ETS (European Union Allowances) were to be eligible to be used for compliance under the AUS CPM from July 2015 until July 2018 ('one-way link'). From 1 July 2018, a two-way link would be put in place, with mutual recognition of carbon units between the two ETSs.[46]

The CPM also allowed for the use of units from credible international carbon markets. However, the most important feature was the negotiated linking with the EU ETS, initially unilateral, which was supposed to commence in 2015.

The ERF/SM currently does not allow for the linking with international units.

## 3.9 Impacts from Cap-and-trade in Australia

*Ex ante* studies of the CPM show that it was estimated to cause a 10 per cent increase in electricity prices, driving significant changes to the energy sector, which would make renewable energy more competitive relative to coal.[47] The Treasury modelling[48] concluded that the impacts on manufacturing output would be small, with some sectors in the economy actually benefiting from the carbon price.[49] While the Treasury modelling stated that some emissions-intensive sectors would require transitional assistance, it did not focus on specific sectors/subsectors of the economy in order to inform the appropriate levels of assistance.

Despite the absence of *ex ante* assessment for specific sectors, in 2011 the Grattan Institute analysed the likely impacts of a carbon price of A$23 rising to A$40 on a number of industry sectors in Australia.[50] It concluded that the industry support in the form of the proposed free allocation of permits to the liquefied natural gas (LNG), coal mining and steel industries was unjustified and costly, putting at risk the environmental efficiency of the scheme and unjustifiably increasing the general costs of carbon reduction elsewhere in the economy, including non-participant sectors and households.[51]

The short period of the scheme's existence did not allow for *ex post* studies on the impacts of the AUS CPM to industry,[52] except for indications that there has been no harm to the overall economy attributable to the CPM.[53]

In terms of its environmental effectiveness, data published on the Quarterly Update of Australia's National Greenhouse Gas Inventory from June 2012 onwards confirms that the CPM had a real potential to reach meaningful emissions reductions. In fact, emissions went down after implementation of the CPM and, inversely, increased after the repeal of the CPM and implementation of the ERF/SM.[54]

In sum, a distinct hierarchy of compliance with sustainability criteria can be detected for former and current market-based climate policy approaches in Australia. While the CPRS clearly had the best design, the CPM also did well and even delivered some promising results in the short period of its efficacy. The ERF/SM, however, cannot be considered a sustainable climate policy choice for Australia.

## 4 THE POLITICS OF CAP-AND-TRADE IN AUSTRALIA

### 4.1 Voters: The Australian Public

The Australian voters' opinion on climate policy has had its ups and downs.[55] In the early 2000s, basically, there was strong support for climate action driven, for example, by the 2007 Australian bushfires as well as international events such as the Al Gore movie *An Inconvenient Truth*. From 2008 onwards, however, attention switched to the global financial crisis and potential additional costs caused by ambitious climate policy.

Australian voters' position on carbon-cap-and-trade is mainly determined by the perception of the pricing and revenue recycling design elements.[56] Early opinion polls show broad support for pricing schemes, which formed the background of former Prime Minister Rudd's proactive strategy on climate change and the CPRS as well as Prime Minister Gillard's CPM. The later public resistance was almost entirely formed by a campaign of the Coalition opposition under the Abbott leadership, terming the CPM a tax and emphasising possible increases in households' costs. Mainly over this issue, in 2013 Australians voted the Gillard government out of office and replaced it by Abbot. It added to the problem, that the revenue recycling was not as easily understood as the pricing itself. The continuing political quarrelling over climate and energy policy, current Prime Minister Turnbull's broken promise to tackle the issue seriously, and the still-sticking Abbott tax tag has left a continuing distrust in Australia's governments' will and skill to deal with climate change.

## 4.2 Interest Groups: Australian Environmental Organisations, Industry Groups and Labour Unions

Major interest groups involved in the debate about cap-and-trade in Australia have been environmental organisations, industry groups and labour unions.

Environmental non-governmental organisations (NGO) have quite generally been supportive of GHG pricing for over a decade now.[57] They mainly value the absolute limit to emissions. Targets in line with the climate protection necessities, comprehensive coverage, and limits to offsets and free allocation have been major design requirements. Still, NGOs opposed Rudd's CPRS because of hopes for an even more stringent scheme. Gillard's CPM, in turn, gained more support, mainly due to the eventual complete CPRS failure. This support for ambitious cap-and-trade continues to date. However, NGOs have changed their strategy from fierce political lobbying to a more cooperative strategy, institutionalised, for example, by the Australian Climate Roundtable. Hopes are that this cooperation with major business associations and labour unions will open more doors for a new cap-and-trade scheme in Australia than continued engagement in the political battles between government and opposition parties. The now increased consensus amongst NGOs on cap-and-trade might also add to NGOs' political power. Still, NGOs' political influence on a coalition government is much smaller than on a Labor government, mainly due to limited personal relationships. Not least, compared to industry, Australian NGOs still lack financial and personnel resources. However, industry has lost some of its influence due to a growing gap between individual sectors' positions.

Major parts of Australian industry have quite substantially changed their position on cap-and-trade.[58] However, the extracting industry to this day continues to doubt climate science and opposes cap-and-trade, mainly because of the threat to coal use and exports. The manufacturing sector, on the other hand, had shown some level of openness before 2010, then joined the extracting industry in its determined opposition, but recently have become more open again. While international competitiveness remains its major concern, above all else, manufacturing craves energy and climate policy certainty for long-term investment decisions. Coverage of the industry, transport, and household sectors, international linkages, and a phase-in similar to the CPRS would be design elements most appreciated. Genuine support for cap-and-trade continues to be provided by the service sector, banks, insurance companies and renewable energy companies. Australian labour unions, even including mining sector unions, have also been long-time supporters of cap-and-trade.

### 4.3 Bureaucracy: the Department of Environment and Energy

Civil servants in Australia's ministries below the minister level, more than in other countries, tend to value independence from political parties' ideologies and interest groups' opinions. Still, officials in the Department of the Environment and Energy have an above-average motivation to protect the environment in general and the global climate in particular.[59] Cap-and-trade, though initially seen with some scepticism, has gained support within the Department at the working level and is now, even after the failure of the CPRS and CPM and their political legacy, considered to be the most promising approach to domestic and international climate policy.

The Department's political influence mainly stems from its officials' technical expertise in the respective matters and their capability of designing and refining policies. This expertise has been acquired over a period of more than 15 years of working on cap-and-trade, and, as a consequence, despite the usual personnel shuffling, the Department still disposes of ample policy design know-how, elaborated policy proposals, and supporting data. However, especially in these times of the particular political legacy of cap-and-trade, lower-level civil servants have limited influence on really driving what ends up being a very political decision made by the governing parties and their cabinets.

## 4.4 Politicians: The Coalition, the Labor Party and the Greens

Despite the differences in the positions described above, the major political battle has been fought between political parties.[60]

Supported mainly by the extracting industry, the conservative-liberal Australian Coalition parties have positioned themselves as opponents of ambitious cap-and-trade. This results partly from an industry-friendly party ideology, but mainly from tactical planning in the campaign leading to the 2013 elections. In order to overcome the Gillard Government, Abbott went on a crusade against the CPM with the tax tag as its winning weapon. Pounding the tax drum all along, eventually he gained the support of the majority of the Australian electorate and was elected prime minister. And immediately he got rid of the CPM, replacing it with the ERF/SM. To this date, even under the Turnbull Government, there has been no intention to open a new discussion on a real cap-and-trade scheme.

The Labor Party continues to support cap-and-trade. However, the Rudd Government at the time missed its chance to implement the CPRS, mainly due to unnecessary political manoeuvring at times when public support for climate action was the strongest and industry resistance the least. And while Gillard revived the idea, she also fed Abbott's campaign by publicly admitting that the CPM, with its initially fixed price, could be termed a tax. Labor's current policy proposal builds on a sectoral approach and a two-step phase-in of cap-and-trade, which starts with an intensity baseline-and-credits scheme and then moves on to a comprehensive full-fledged cap-and-trade program.

The Green Party, though actually in favour of ambitious cap-and-trade, had its devastating moment when, backed by more fundamental NGOs at the time, they voted against Rudd's CPRS on the basis of a design critique in detail. Having learned the political lesson, they later supported Gillard's CPM. Currently, the Greens still favour an ambitious cap-and-trade program as the best climate policy option to deliver the emissions reductions necessary to meet Australia's Intended Nationally Determined Contributions (INDC) target.

In sum, while the fierce battle over cap-and-trade between Australia's political parties continues, the opposing forces are now limited to the Coalition parties and extracting industries. Support, on the other hand, comes from the Labor and Green Parties, manufacturing and other businesses, labour unions and NGOs. While a newly elected Labor-led government could certainly bring about major change after the 2019 elections, the Coalition's margin for manoeuvring towards a new cap-and-trade scheme seems limited.

## 5 CONCLUSIONS

A number of uncoordinated policies have been debated, introduced and dismantled over the past ten years in Australia. Inconsistent approaches towards climate action were fuelled by a toxic political debate, not only around the ideal policy mechanism but around the science of climate change itself.

The CPRS and the CPM were to a great extent compliant with the sustainability criteria for a cap-and-trade and would be improved over time, however the schemes lacked political support. The ERF/SM, while politically feasible, has not been capable of generating meaningful emissions reductions and encouraging the transition to a low-emissions economy in Australia.

Of the three schemes that were compared, the ERF/SM is by far the worst in terms of compliance with sustainability criteria. Some of the most concerning features are the incapacity to raise revenue (exclusively relying on substantial government funding)[61] and the lack of stringency of the baselines under the SM.

Despite the complex political landscape, Australia will be increasingly pressured to fulfil its commitments under the Paris Agreement and it is clear that the measures currently in place are insufficient. The lack of political certainty has harmed investment and now some new pro-active dynamics are starting to be shaped in the business community. There is scope, therefore, for a cautiously optimistic view that, under certainly necessary new political leadership, 'the recent tarnishing of cap-and-trade . . . will . . . turn out to be a temporary departure from a long-term trend of increasing reliance on market-based environmental policy instruments'.[62]

## NOTES

* Discussion Paper presented at the 18th Global Conference on Environmental Taxation (GCET18), University of Arizona, James E. Rogers College of Law, Tucson, 27–29 September 2017. The research was generously supported by the University of Newcastle 2017 International Research Visiting Fellowship (IRVF). Special thanks also go to the interview partners in Australia and Mashifu Noguchi for helping with the data processing.

1. Alfred Endres, *Environmental Economics: Theory and Policy* (Cambridge University Press, 2011).
2. Rudolph, S, Lenz, C, Lerch, A and Volmert, B, 'Towards Sustainable Carbon Markets'. In: Kreiser, L, Yábar Sterling, A, Herrera, P, Milne, JE and Ashiabor, H (eds), *Carbon Pricing, Growth and the Environment – Critical Issues in Environmental Taxation XI* (Cheltenham, UK/Northampton, US: Edward Elgar Publishing, 2012) 167–83.
3. Paris Agreement, Art 3.
4. Rudolph S, Lerch, A and Kawakatsu, T, 'Developing the North American Carbon

Market – Prospects for Sustainable Linking'. In: Weishaar, S, Kreiser, L, Milne, JE, Ashiabor, H and Mehling, M (eds), *The Green Market Transition: Carbon Taxes, Energy Subsidies and Smart Instrument Mixes – Critical Issues in Environmental Taxation Volume XIX* (Cheltenham, UK/Northampton, US: Edward Elgar Publishing, 2017) 209–30.

5. ICAP, *Emissions Trading Worldwide: Status Report 2017* (Berlin: ICAP).
6. Kirchgässner, G and Schneider, F, 'On the Political Economy of Environmental Policy' (2003) *Public Choice* 115(3), 369–96.
7. Hahn, RW, 'Jobs and Environmental Quality – Some Implications for Instrument Choice' (1987) *Policy Sciences* 20(4), 289–306, 289.
8. Schneider, F, Kollmann, A and Reichl, J, *Political Economy and Instruments of Environmental Politics* (Cambridge, MA: MIT Press, 2015).
9. Carbon Pollution Reduction Scheme Bill 2009 (Cth) ('CPRS Bill').
10. Garnaut, R, *The Garnaut Climate Change Review: Final Report*, (New York: Cambridge University Press, 2008) 101.
11. Department of Climate Change, 'Carbon Pollution Reduction Scheme: Green Paper' (2008); Commonwealth of Australia, 'Australia's Low Pollution Future: The Economics of Climate Change Mitigation' (2008); Department of Climate Change, 'Climate Change Carbon Pollution Reduction Scheme: White Paper' (2008).
12. See Sopher, P, Mansell, A and Munnings, C, *Australia* (EDF IETA, 2014).
13. Ibid.
14. Australian Government, 'Securing a Clean Energy Future: The Australian Government's Climate Change Plan' (2011).
15. Clean Energy Act 2011 (Cth) ('CE Act'); Clean Energy Regulator Act 2011 (Cth); Climate Change Authority Act 2011 (Cth); Australian National Registry of Emissions Units Act 2011 (Cth); Clean Energy (Charges – Customs) Act 2011 (Cth); Clean Energy (Charges – Excise) Act 2011 (Cth); Clean Energy (Consequential Amendments) Act 2011 (Cth); Clean Energy (Household Assistance Amendments) Act 2011 (Cth); Clean Energy (Unit Issue Charge – Auctions) Act 2011 (Cth); Clean Energy (Unit Issue Charge – Fixed Charge) Act 2011 (Cth); Clean Energy (Unit Shortfall Charge – General) Act 2011 (Cth); Clean Energy (Tax Laws Amendments) Act 2011 (Cth).
16. Clean Energy Legislation (Carbon Tax Repeal) Act 2014 (Cth) sch 1 pt 1.
17. See Section 3.2 of this chapter.
18. CFI Act pt 2A; CER, *Understanding Contracts* (4 April 2016) <http://www.cleanenergyregulator.gov.au/ERF/Want-to-participate-in-the-Emissions-Reduction-Fund/Step-2-Contracts-and-auctions/understanding-contracts> ('*ERF Contracts*').
19. Ibid s 20F.
20. See DoE, *Reducing Australia's Emissions* (2014) <https://www.environment.gov.au/system/files/resources/5acdfbf8-8ced-4c54-a61a-c06cb31d1e88/files/reducing-australias-emissions.pdf>.
21. Ibid 1.
22. Rudolph et al., above n 4.
23. CPRS White paper, xxxviii.
24. Explanatory Memorandum, 32.
25. CE Act s 20(4); Clean Energy Regulator, *LEPID for the 2013–14 Financial Year* (30 June 2015) Australian Government <http://www.cleanenergyregulator.gov.au/Infohub/CPM/Liable-Entities-Public-Information-Database/LEPID-for-the-2013-14-financial-year>.
26. Rail, domestic aviation and shipping.
27. The AUS CPM covered exclusively $CO_2$, $CH_4$, $N_2O$ and PFCs from aluminium smelting.
28. Rudolph et al., above n 4.
29. Ibid s 100(7).
30. Ibid s 14.
31. *CE Act* s 14(2).
32. Australian Government, 'Securing a Clean Energy Future: The Australian Government's Climate Change Plan' (2011).

33. CFI Act pt 2A; *ERF Contracts*', above n 18.
34. Explanatory Memorandum, Clean Energy Bill 2011 (Cth) 13.
35. Rudolph et al., above n 4.
36. Explanatory Memorandum, CPRS Bill, 2.39.
37. Ibid ss 121, 123A (8).
38. Rudolph et al., above n 4.
39. CE Act s 100(1).
40. Explanatory Memorandum, Clean Energy Bill 2011 (Cth) 32.
41. Explanatory Memorandum, Clean Energy Legislation Amendment (International Emissions Trading and Other Measures) Bill 2012 and related Bills (Cth) 4.
42. Rudolph et al., above n 4.
43. Clean Energy (Unit Shortfall Charge—General) Act 2011 (Cth) s 8(3)(a).
44. Ibid s 8(3)(b).
45. CE Act s 135(1)(a)(b).
46. Australian Government, 'Australia and European Commission agree on pathway towards fully linking Emissions Trading Systems' (2012) <http://www.climatechange.gov.au/en/media/whats-new/linking-ets.aspx>.
47. Also see Commonwealth of Australia, 'Strong Growth, Low Pollution: Modelling a Carbon Price. Update' (2011) 96. The estimates were confirmed in the modelling update.
48. Commonwealth of Australia, 'Strong Growth, Low Pollution: Modelling a Carbon Price' (2011).
49. Ibid 98, 113.
50. Wood, T and Edis, T, *New Protectionism under Carbon Pricing: Case Studies of LNG, Coal Mining and Steel Sectors* (Grattan Institute, 2011) 26.
51. Ibid.
52. See O'Gorman, M and Jotzo, F, *Impact of the Carbon Price on Australia's Electricity Demand, Supply and Emissions* (Centre for Climate Economics & Policy, 2014).
53. Ibid. See also Twomey, P, *Obituary: The Carbon Price* (UNSW Australia) <http://newsroom.unsw.edu.au/news/business/obituary-carbon-price>.
54. See Australian Government Department of the Environment and Energy, 'Publications and Resources' (2017) *Quarterly Updates of Australia's National Greenhouse Gas Inventory* <http://www.environment.gov.au/climate-change/greenhouse-gas-measurement/publications>.
55. Climate Institute 2017: personal interview with a Climate Institute representative, February 2017; Research A 2017: personal interview with a researcher, February 2017.
56. eNGO B 2017: personal interview with an environmental NGO representative, February 2017; Political Party A, B 2017: personal interviews with political party representatives, February 2017; Research B, C, D 2017: personal interviews with researchers, February 2017.
57. Climate Institute 2017: personal interview with a Climate Institute representative, February 2017; eNGO A, B 2017: personal interviews with environmental NGO representatives, February 2017.
58. Steel Industry 2017: personal interview with a steel industry representative, February 2017; MCA 2017: personal interview with a representative of the Minerals Council of Australia, February 2017; Natural Resources Industry 2017: personal interview with a representative from a natural resources company, February 2017; Conroy 2017: personal interview with Patrick Conroy (MP), February 2017.
59. eNGO A, B 2017: personal interviews with environmental NGO representatives, February 2017; State Government Department 2017: personal interview with a State Government department representative, February 2017.
60. Political Party A, B 2017: personal interviews with political party representatives, February 2017; Conroy 2017: personal interview with Patrick Conroy (MP), February 2017.

61. See, e.g. Harry Clarke, Iain Fraser and Robert Waschik, 'How much abatement will Australia's Emissions Reduction Fund buy?' (Crawford School of Public Policy, 2014).
62. Schmalensee, R and Stavins, R, 'Lessons Learned from Three Decades of Experience with Cap-and-Trade' (2015) Harvard Kennedy School of Government Faculty Research Working Paper RWP15-069, p 19.

# 5. How market-based emissions reduction mechanisms affect private property in Australia

**Vanessa Johnston**

## 1 INTRODUCTION

Climate change is widely accepted to pose a range of risks and challenges for Australia. As a result, Federal, State/Territory, and local governments are taking action to both mitigate, and adapt to, the consequences of climate change. It is well recognised that adapting to climate change is changing the way that land can be used, particularly in relation to coastal and low-lying land that is, or will be, vulnerable to sea-level rise, flood, storm surges or inundation expected from climate change. However, it is less obvious that activities currently undertaken to avoid climate change, particularly to mitigate or offset greenhouse gas (GHG) emissions, can also affect land use. In this context, command-and-control and market mechanisms implemented by the State to encourage entities to mitigate GHG emissions can significantly affect the rights and obligations that entities can exercise over affected private property.

This chapter explores two overarching issues of how activities undertaken to mitigate climate change are affecting land use in Australia. First, what are the nature and scope, and the grounds for exercising rights over private property under Australian law in the context of climate change? Second, how is use and enjoyment of private property affected by Australia's flagship mechanism to encourage GHG emissions mitigation, the Emissions Reduction Fund (ERF)?

## 2 THE CONCEPT OF PROPERTY AND THE NATURE OF PRIVATE PROPERTY RIGHTS

Entities can hold a variety of estates or interests that entitle them to exercise certain rights over land under Australian law. Based on historical English feudal structures of land holding, Australia's system of land ownership divides rights in land between the State and private entities.[1] The State holds and retains ultimate or radical title to land that affords rights associated with both use and enjoyment of land as well as unique rights to grant further estates and interests in land, exercise reserved rights over resources, and compulsorily acquire land on terms set out in the Australian Constitution.[2] The absolute title granted in land by the State to private entities gives rise to a 'bundle of rights' dependent on the nature of the estate or interest held. In general, rights that private entities can exercise over land include those of use, possession, alienation, disposal, and the right to exclude.[3] Owners of land (i.e. the registered proprietor/holder of a freehold estate) generally exercise all of these rights.[4] Tenants (holders of a leasehold estate) have the right to exclusive possession and use, but have limited rights of alienation and disposal.[5] Non-possessory proprietary interests in land only confer rights of use and/or possession in limited circumstances; mortgagees (security interests) have rights to take possession of land after default of the mortgagor, and also to dispose of and/or alienate land as conferred by statute.[6] Beneficiaries of an easement or profit a prendre (servitude interests) can exercise limited use rights (e.g. right of way or access) in accordance with the interest held.[7] Licensees can enjoy non-exclusive rights to use and occupy land pursuant to a contractual licence.

It is widely accepted that the identity of, and relationship between, subjects and objects of property are not closed.[8] Nevertheless, in Western legal systems State sanction or recognition of subjects, objects, and rights over property is essential,[9] and must exist at all times.[10] Consequently, the State (either via the parliament or judiciary) plays an important role in shaping the sanctioned and recognised classes of property, and the rights that private entities may exercise in relation to that property. In Australia, the nature and scope of rights exerciseable over land generally align with the classical liberal view that an entity is entitled to exercise property rights in its own self-interest as a privilege arising from the input of labour or capital.[11] This dominant view succeeds over the alternative views of social obligation, under which property rights granted to an entity over land come hand-in-hand with responsibilities and obligations of stewardship.[12]

In the context of climate change, the latter view is more likely to contribute to desired environmental outcomes, whereas the former arguably facilitates behaviours that contribute to climate change.[13] Regulatory measures that encourage mitigation of GHG emissions by encouraging certain activities and discouraging others, in relation to land at least, create a connection between dominant and alternative views on exercising rights over private property as stated above. Effectively, regulatory measures to encourage uptake of GHG emissions mitigation activities can change the terms on which the State is prepared to sanction a private entity exercising rights over land in light of climate change. The link between these regulatory measures relating and the way an entity can exercise rights over property is closest where the regulatory measures directly encourage or discourage land-use activities – as would occur in the agricultural and forestry sectors. However, this can also occur where the GHG emissions mitigation activity is carried out or requires the use of specific land.

## 3 REGULATORY CONTEXT AND BACKGROUND ON LAND USE IN AUSTRALIA

Australia is obliged to contribute to global efforts to mitigate climate change under international law. Relative to 2005 levels, Australia has agreed to reduced national GHG emissions by 13 per cent by 2020 (Kyoto Protocol),[14] increasing to 26–28 per cent by 2030 (Paris Agreement).[15] Australia's commitment under the Kyoto Protocol for 2020 is reflected in the national climate change policy, the *Direct Action Plan*,[16] which supports a range of regulatory measures intended to encourage mitigation of GHG emissions. The flagship mechanism of the *Direct Action Plan* is the ERF, established by the Carbon Credits (Carbon Farming Initiative) Act 2011 (Cth) ('Carbon Credits Act'). The government pays entities from the ERF to undertake eligible projects to mitigate GHG emissions (total funding of AUD$2.55 billion).[17]

An entity that has the legal right to carry out a project to mitigate GHG emissions[18] receives funding from the ERF after successful completion of a two-stage process set out in the Carbon Credits Act. In the first stage, an entity applies to the governing Clean Energy Regulator to undertake the 'eligible offsets project', which must meet specific prescribed eligibility criteria.[19] These criteria include that project activities comply with existing 'methodology determinations', which are designed to assist entities to calculate emissions savings for the project and therefore the quantity of carbon credits sought from the Clean Energy Regulator. Types of activities that may be implemented as an eligible offsets project vary widely, relating to waste treatment, piggeries, commercial buildings, land and sea transport, revegetation, avoided clearing, native forest management, savannah burning, and soil carbon sequestration.[20] This extensive list ensures that activities in key sectors that contribute to and affect Australia's GHG emissions, as listed in Table 5.1, are encouraged to mitigate GHG emissions with financial assistance from the ERF. If the entity and its

*Table 5.1 Land use/GHG emissions by economic sector*

| Economic sector | Percentage of Australia's GHG emissions (2015)* | Percentage of Australia's total land area (7,687,124 km²)** |
|---|---|---|
| Stationary energy | 53% | 0.04% (311 km²) |
| Transport | 18% | 0.014% (1,080 km²) |
| Fugitive emissions (fuel) | 8% | NA |
| Industrial processes | 6% | 0.009% (704 km²) |
| Agriculture | 13% | 58% (4,454,749 km²) |
| Waste | 2% | 0.004% (307 km²) |
| Forestry (LULUCF) | −4% | 11.8% (906,434 km²) |

*Notes:*
* Australian Government (2017), *National Inventory Report 2015* (Report, Volume 1) [table 2.1; figure 2.2] 32, 37. Australian Government (2016), 'Australia's Emissions Projections 2016', 25.
** Australian Bureau of Agricultural Resource Economics and Sciences (2016), 'Land Use of Australia Dataset 2010–11', *Land Use Data Download; Department of Agriculture*, http://www.agriculture.gov.au/abares/aclump/land-use/data-download.

*Sources:* Australian Government (2017), *National Inventory Report 2015* (Report, Volume 1) [table 2.1; figure 2.2] 32, 37; Australian Government (2016), 'Australia's Emissions Projections 2016', 25; Australian Bureau of Agricultural Resource Economics and Sciences (2016), 'Land Use of Australia Dataset 2010-11' *Land Use Data Download; Department of Agriculture*, < http://www.agriculture.gov.au/abares/aclump/land-use/data-download>.

project satisfies these requirements, the Clean Energy Regulator will issue Australian Carbon Credit Units ('carbon credits') to that project – one for each tonne of GHG emissions that will be saved or avoided over the project term.[21] While the ERF mechanism ultimately pays entities directly to mitigate GHG emissions (i.e. direct financial assistance), stage one of the eligibility process operates similarly to a baseline-and-credit emissions trading scheme[22] because the Clean Energy Regulator issues carbon credits for GHG emissions savings below the entity's historical emissions baseline. However, currently the ERF is a non-trading scheme: there is no secondary market in which entities can buy, sell or trade in carbon credits.[23]

In the second stage, successful entities may sell their carbon credits to the Australian government in a reverse auction.[24] Reverse auctions encourage mitigation of GHG emissions at least cost[25] to maximise the amount of GHG emissions savings that can be purchased from entities in light of Australia's mitigation targets. Successful entities enter into a sale contract with the Clean Energy Regulator (as an agent of the Australian government),[26] which sets out obligations regarding the project and delivery of carbon credits over the project term (on average 7–25 years).[27] There have been five auctions held under the Carbon Credits Act to date. Consequently, the Clean Energy Regulator has entered into contracts for 435 mitigation projects, which will deliver GHG emissions savings of 189 million tonnes, at an average price of AUD$11.83 per tonne[28] (cost to date AUD$2.23 billion).[29] Compared to economic carbon pricing models, and the price paid for GHG emissions in other mitigation schemes, the price under the ERF is relatively low.[30] Accordingly, only projects with low implementation and operational costs have been successful at ERF auctions. The most successful projects at these auctions, by quantity, are projects implemented pursuant to vegetation methodologies (373 projects), followed by landfill/waste (174 projects) and agriculture (40 projects).[31] The nature of activities associated with projects in these areas suggests entities are choosing to use land in specific ways to capitalise on funding opportunities under the ERF. However, as shown in Table 5.1, comparing data from Australia's most recent National Greenhouse Gas Inventory Report to land-mapping data produced by the Australian Bureau of Resource Economics shows little correlation between sectoral contributions to GHG emissions and land use.

Nonetheless, changes to these activities could have substantial impacts on GHG emissions reduction efforts and land use in Australia in the future. For example, changes to activities in the stationary energy sector will have little impact on land use, and the exercise of private property rights over land used for stationary energy activities (0.04 per cent), but could have significant impact on GHG emissions (53 per cent). In contrast, changes to agricultural activities could have a moderate impact on GHG emissions (13 per cent), but significant impact on the exercise of private property rights over land (58 per cent). As a 'sector' land-use, land-use change, and forestry (LULUCF) activities result in negative GHG emissions (−4 per cent)[32] because they involve carbon sequestration (vegetation), and therefore contribute to Australia's carbon sinks. On this basis, changes to LULUCF activities can significantly offset national GHG emissions over the long term. While changes to private property rights exercised over land used for LULUCF activities will affect a moderate proportion of Australian land (11.8 per cent), the long duration of LULUCF projects means that the potential impact on private property rights is substantial. These possible impacts on both land use and GHG emissions indicate the importance of understanding the underlying relationship between land use and GHG emissions in Australia, so that both objectives can be achieved in the future.

## 4 HOW DOES THE ERF ENCOURAGE LAND-USE CHANGE?

In circumstances where the ERF makes it commercially viable for an entity to undertake a project to mitigate or offset GHG emissions that are eligible for assistance under the Carbon Credits Act, the ERF can influence land-use decisions made by an entity with the legal right to use land.[33] At an average price of AUD$11.83 per tonne, the variety of feasible GHG emissions mitigation or avoidance activities that entities implement in return for carbon credits may be limited.[34] For example, in the largest class of successful ERF projects by method (vegetation methodology – 373 projects) approved activities vary widely, ranging from those that merely protect existing land use, to those that involve new or altered land use. To date, there are 60 projects registered to receive carbon credits for sustaining existing land uses, namely by avoiding deforestation or regrowth clearing.[35] Maintaining existing land use is important: in 2012, for example, approximately 10 per cent of Australia's GHG emissions[36] arose in the process of converting forest land to 'other uses'.[37] Consequently, the ERF encourages entities to maintain existing privately owned forests (currently 333.94 square kilometres)[38] in their current state, thereby enhancing GHG emissions savings from this source. In addition, the ERF encourages entities to convert land into forests to increase the opportunities for offsetting GHG emissions. Approved activities also include those which encourage new uses of land. To date, 85 projects have received carbon credits for revegetation or new plantings around Australia (equivalent to almost 23 per cent of all successful vegetation projects).[39] GHG emissions offset by new forests have increased from 7.3 Mt to 10.4 Mt since the ERF commenced in 2014.[40]

These figures alone are no more indicative of the fact that vegetation projects that strengthen Australia's carbon sinks are financially viable at the price currently paid for carbon credits under the ERF. Nonetheless, considering the entities involved in these vegetation projects suggests that land is being acquired to specifically generate carbon credits and is therefore increasing the percentage of Australian land used for vegetation purposes, which has flow-on effects for land use and property rights as explained below. Businesses that offer emissions-intensive products or services often provide customers with the opportunity to offset GHG emissions from those activities: airlines (e.g. QANTAS, Virgin Australia, TigerAir), manufacturers (AustralBricks, Australian Paper), energy suppliers (e.g. Energy Australia, Castrol), and even travel companies (e.g. Intrepid Travel) have achieved 'carbon neutral' certification. Conditions attached to such certification, as illustrated by the Australian government's 'Carbon Neutral

Program',[41] include compliance with 'best practice' guidelines to measure, monitor, and reduce GHG emissions, independent audit, and public scrutiny.[42] While these conditions enhance the outward perception of certified companies as 'environmentally conscious', participating in carbon-neutral programs also contributes to environmental social governance reporting between corporations and their shareholders.[43] Consequently, there is increasing demand for carbon offset providers – entities who implement projects to generate carbon credits under the ERF on behalf of corporate and individual clients that provide offset opportunities to their customers. In Australia a carbon credit is a financial product under the Corporations Act 2001 (Cth)[44] and therefore entities that deal with, or provide advice in respect to, carbon credits (such as carbon offset programs) must hold an Australian Financial Services Licence and comply with statutory reporting and disclosure requirements.[45]

The potential impact that the carbon offset market has on land use is illustrated by the activities of established carbon offset providers, including Greenfleet, auscarbon, and CarbonNeutral. Greenfleet is a not-for-profit organisation that has provided offsets for transport activities specifically through vegetation projects. Since 1997 Greenfleet has invested in planting projects in 475 forests across Australia and New Zealand, comprising more than 8.9 million native trees.[46] Notably, while Greenfleet is the named contractor for environmental plantings projects in Victoria and New South Wales under the ERF,[47] they also actively market to landowners, seeking opportunities to revegetate private land at 'no cost'.[48] CarbonNeutral and auscarbon are both involved in plantings projects in the YarraYarra Biodiversity Corridor in Western Australia. Since 2007, auscarbon has established 13 planting projects designed to co-exist with other agricultural uses.[49] It is the contractor for five current projects under the ERF.[50] CarbonNeutral also provides carbon offset opportunities for clients with emission-intensive businesses – as a partner that provides the 'Future Forests' program for stationary energy retailer AGL Energy,[51] and as a means by which construction machinery manufacturer Hitachi offsets GHG emissions from 550 corporate fleet vehicles and air travel within its corporate group.[52]

In light of the above, it is clear that market mechanisms such as the ERF are directly encouraging land-use projects that avoid/offset GHG emissions and indirectly creating a demand for carbon offsets, both having real and tangible impacts on land use in Australia. This process increases the proportion of land used for vegetation/forestry activities in Australia[53] and enhances the impact of any changes to rights exercised over land used to mitigate GHG emissions under the Carbon Credits Act. These trends are important in the context of how private property rights over forestry and agricultural land may be restricted for the duration of the contracted mitigation project, indirectly 'locking in' long-term land use.

## 5 HOW DOES THE ERF AFFECT PRIVATE PROPERTY RIGHTS?

As discussed above, the right to use land is one of the essential rights associated with holding an interest in property. It may be exercised by holders of many proprietary interests (e.g. freehold and leasehold estates), or pursuant to personal interests arising under contractual licences of occupation. By enacting the Carbon Credits Act, the Australian government has changed the terms on which it sanctions the exercise of private property rights over land affected by the Act. In other words, provisions of the Carbon Credits Act can restrict (conditionally or unconditionally) an entity's ability to use land connected to GHG emissions mitigation and/or avoidance activities funded by the ERF. The impact of the Carbon Credits Act on private property rights differs according to the activities carried out on the land. In particular, the Clean Energy Regulator can impose both positive conditions and negative restrictions on land used for sequestration offsets projects (vegetation or soil)[54] via 'carbon maintenance obligations' (CMO), which may affect land long after the project term has expired.

Part 8 of the Carbon Credits Act empowers the Clean Energy Regulator to impose CMOs on contractors and their projects in numerous circumstances. In relation to project eligibility, the Clean Energy Regulator may impose a CMO if carbon credits are found to have been issued based on false or misleading information (s 88) or if the project's 'eligible offsets project' has been revoked (s 89). In relation to eligible sequestration offset projects, the Clean Energy Regulator can also impose a CMO if there is a significant reversal of sequestration due to either (i) a man-made disturbance or conduct outside the control of the contractor (s 90) or (ii) a natural disturbance which is not mitigated by the contractor within a reasonable time (s 91). If a CMO is imposed on a project, the Clean Energy Regulator must take all reasonable steps to provide copies of the CMO declaration to the contractor, any person with an identified interest in the land,[55] and to the relevant land registration official (Registrar of Titles).[56] In Australia, the Registrar of Titles in each State and Territory has the discretion, but not an obligation, to note the existence of the CMO on the title of the affected land.[57] Importantly, CMOs will remain in force until the Clean Energy Regulator declares that the contractor has discharged the required conditions/penalties, for the duration of the project, or the duration of the 'permanence period', whichever occurs first.[58]

The Clean Energy Regulator imposes permanence periods on all sequestration offset projects to ensure achievement of GHG emissions savings relative to projects that do not rely on storage of carbon.[59] Permanence periods increase the duration of time in which project activities must be carried out to ensure GHG emissions mitigation, to manage the risk of exposure that these projects have to carbon reversal caused by anthropogenic and natural events such as land-use change, land clearing, timber harvesting, bushfire, disease, pests, or vegetation death.[60] Generally, projects involving sequestration (timber) and revegetation (plantings) are subject to permanence periods of 100 years, with lesser periods of 25 years applicable to projects involving reforestation (planting) and sequestration (soil).[61] There are currently 417 vegetation and agricultural projects recorded on the ERF register as being subject to permanence periods, 260 of which have been allocated the maximum permanence period of 100 years.[62] Combined, these projects are implemented on several thousand individual land parcels.[63]

Permanence periods enhance the impact of both positive and negative conditions imposed on the exercise of private property rights on the project land (as a whole, or in part) by the Clean Energy Regulator pursuant to the CMO.[64] More specifically, CMOs may identify 'permitted carbon activities' for land to minimise risks or address carbon reversal.[65] In this case, the Clean Energy Regulator can stipulate the area(s) of land, manner, time(s), and duration for the permitted carbon activity, including the persons involved.[66] While the contractor is the most relevant party to discharge the requirements of the CMO, the Carbon Credits Act imposes these obligations on both owners and occupiers of the relevant land.[67] The Clean Energy Regulator can therefore demand compliance with the CMO from the holders of any proprietary estates and interests in land that confer ownership/occupation rights (e.g. freehold estates, leasehold estates, mortgages, etc) in addition, or as an alternative to, the named contractor[68] until the expiration of the project's permanence period (100 years).

The potential for restrictions or obligations arising from land use and/or eligibility under the ERF for the duration of permanence periods to affect non-contracting entities should not be underestimated. The carbon contact between the initial project entity and the Clean Energy Regulator gives rise to personal obligations that will only be enforceable against contracting parties unless legally assigned to others.[69] However, if a subsequent estate or interest holder acquires its estate or interest from the contractor with notice of the carbon contract or ERF status of the activities carried out on the land, it may become liable to fulfil contractual and statutory provisions as protected by the Carbon Credits Act.[70] To the extent that eligible sequestration projects also fulfil the requirements of 'forest carbon' projects under State and Territory legislation, GHG emissions mitigation activities may give rise to a separate statutory interest in land.[71] In these circumstances, carbon contracts may affect not only private property rights of use and possession, but also rights of alienation and disposal by requiring the forest/carbon right holder to provide consent before a freehold or leasehold estate holder can assign or transfer its interest in land.

## 6 COMMENT AND CONCLUSION

Based on the experienced and likely outcomes of the ERF as administered through the Carbon Credits Act, it is clear that market mechanisms enacted to encourage GHG emissions mitigation activities both indirectly influences land use choices in Australia and restrict how the holders of estates and interests in land exercise private property rights associated with their estate or interest. Notably, decisions made by entities to participate in the ERF now can have an ongoing impact on land use and the exercise of private property rights in the future. Decisions to capitalise on ERF opportunities for forestry and agricultural activities involving the sequestration of carbon in vegetation and soil raise the possibility of permanence periods, and 'locking in' of land use for up to 100 years – far longer than the initial contractor is likely to remain in control of, or associated with, the relevant land. Case studies from carbon offset providers also suggest that land is sometimes being put to multiple uses – the co-existence of vegetation activities on existing agricultural land effectively encumbering private landholdings with additional restrictions imposed by the Carbon Credits Act.

Market mechanisms encourage behavioural changes that result in fewer GHG emissions. In relation to climate change objectives, this outcome should always be encouraged. However, the lack of transparency and awareness about the connection between these new behaviours and the impact on long-term land use and private property rights in Australia is problematic. The infancy of the Carbon Credits Act and the carbon sequestration activities funded by the ERF means that it is uncertain how contractors and estate/interest holders in land will deal with the period of time between the expiration of the project (25 years) and permanence periods (up to 100 years). Arguably, the landowner may become burdened with the risk of maintaining the existing land use until the expiration of the permanence period at 'no cost' (as promised by carbon offset providers) but also perhaps 'no gain'. The extent to which private contractual arrangements[72] address this issue is unknown.

Additionally, while GHG emissions mitigation activities to enhance Australia's carbon sinks and sequester carbon in vegetation and soil are clearly beneficial to climate change objectives, these activities do not change the emissions-intensive processes and ingrained social, economic, and political values that undermine Australia's transition towards a low-carbon economy. In the context of land use, these include perceptions and expectations about private property which align with the liberal view that private property is a means of generating wealth, and entities are entitled to exercise private property rights in a self-interested way at all costs. On this basis, the Carbon Credits Act challenges the current understanding of private property under Australian law, and foreshadows the broader impact that regulatory measures will have on the exercise of private property rights within the constraints of a low-carbon economy. It may be that land use and private property rights may need to change to accommodate a new low-carbon economy in Australia. Consequently, both how to bring these indirect impacts to the awareness of private citizens without threatening the momentum of the ERF, as well as how to avoid GHG emissions mitigation incentives disproportionately affecting land use compared to activities in other economic sectors, requires further consideration.

# NOTES

1. In this chapter 'private entities' is used to describe any entity that can hold an estate or interest in land other than the Crown – while this is mainly individuals and corporations, it also extends to public and quasi-public authorities (local councils, utility providers, water authorities and rail authorities).
2. Australian Constitution s 51(xxxi); note that the rights of the State under common law operate parallel to native title rights held by Indigenous Australians, see *Mabo v Queensland* (1992) 175 CLR 1.
3. *Milirrpum v Nabalco Pty Ltd* [1971] 17 FLR 141, 272.
4. Ibid.
5. *Radaich v Smith* (1959) 101 CLR 209; *Lace v Chantler* [1944] KB 368.
6. For example see Transfer of Land Act 1958 (Vic) s 78.
7. *Re Ellenborough Park* [1956] Ch 131; *Clos Farming Estates v Easton* [2002] NSWCA 389.
8. For example *Yanner v Eaton* (1999) 201 CLR 351.
9. Cohen, Felix (1954), 'Dialogue on Private Property', *Rutgers Law Review* 9, 357–87, cf *Mabo* above n 2.
10. Cohen, above n 9, 371–4.
11. France-Hudson, Ben (2017), 'Surprisingly Social: Private Property and Environmental Management', *Journal of Environmental Law* 29 101–27, 105.
12. Ibid, 107; also Alexander, Gregory (2013), 'Ownership and Obligations: The Human Flourishing Theory of Property', *Hong Kong Law Journal* 43, 451–62, 461–2.

13. Babie, Paul (2010), 'Idea, Sovereignty, Eco-colonialism and the Future: Four Reflections on Private Property and Climate Change', *Griffith Law Review* 19, 527–66, 542.
14. Kyoto Protocol to the Framework Convention on Climate Change, opened for signature 16 March 1998, 2303 UNTS 148 (entered into force 16 February 2005) annex B; Australian Government, 'Australia's 2030 Climate Change Target' (Factsheet, 2015) http://www.environment.gov.au/climate-change/publications/factsheet-australias-2030-climate-change-target.
15. UNFCCC, Adoption of the Paris Agreement, Decision CP.21, 21st sess, Agenda Item 4(b), UN Doc FCCC/CP/2015/L.9/Rev.1 (12 December 2015); Commonwealth of Australia, *Australia's Intended Nationally Determined Contribution to a New Climate Change Agreement* (August 2015) http://dfat.gov.au/international-relations/themes/climate-change/submissions/Documents/aus-intended-nationally-determined-cont-new-cc-agreement-aug-2015.pdf.
16. Liberal Party of Australia, *The Coalition's Direct Action Plan* (2011).
17. Australian Government (2014), *Emissions Reduction Fund: White Paper* 8; See also Australian Government (2013), 'Emissions Reduction Fund: Green Paper' (Report). While the available funding proposed at the commencement of the scheme was to cover projects approved between 1 July 2014 and 30 June 2017, this has not yet been increased due to an excess of funding still available as at the 2017 Federal Budget in May 2017, see Australian Treasury (2017), *Budget Papers 2017–18: Portfolio Budget Statements* 178; further Mark Ludlow, 'No top up for Emissions Reduction Fund in May budget' (Australian Financial Review, online, 17 April 2017).
18. Carbon Credits Act s 5 (definition of 'project proponent'). Note that an applicable carbon sequestration holder is now an eligible interest holder under s 43, who must consent to the project as part of the application process; s 28A.
19. Carbon Credits Act s 27. Eligibility criteria cover a wide variety of issues. For discussion of some criteria which affect land-type projects, see Vanessa Johnston (2016), 'Sowing the Seeds of Change: Why Australia's Land Sector Needs a Carbon Price to Encourage Mitigation of GHG Emissions and Promote Sustainable Land Use' in Stoianoff, Natalie P, Larry Kreiser, Bill Butcher, Janet E Milne and Hope Ashiabor (eds), *Critical Issues in Environmental Taxation Vol XVIII: Market Instruments and the Protection of Natural Resources* (Edward Elgar Publishing) 35.
20. Clean Energy Regulator, 'Opportunities for industry', *Emissions Reduction Fund* (22 September 2016) http://www.cleanenergyregulator.gov.au/ERF/Choosing-a-project-type/Opportunities-for-industry; Clean Energy Regulator, 'Opportunities for the land sector', *Emissions Reduction Fund* (22 January 2016) http://www.cleanenergyregulator.gov.au/ERF/Choosing-a-project-type/Opportunities-for-the-land-sector. A complete list of methods established by legislative instruments can be found at www.comlaw.gov.au.
21. Carbon Credits Act s 147.
22. The previous version of the Carbon Credits Act applied only to the agricultural and forestry sectors, enabling these sectors to generate carbon credits which could be surrendered as part of Australia's former 'carbon tax' under the Clean Energy Act 2011 (Cth) (repealed).
23. Clean Energy Regulator (August 2016), 'Market Sounding: Direct Abatement Offers'; Commonwealth of Australia (10 December 2014) 'Request for "Special Review" by the Climate Change Authority', *Climate Change Authority*, http://www.climatechangeauthority.gov.au/files/files/special-review-request.pdf.
24. Carbon Credits Act s 20F.
25. Carbon Credits Act s 20G.
26. Carbon Credits Act s 20B.
27. Carbon Credits Act s 69.
28. Clean Energy Regulator (2017), 'Auction April 2017: Cumulative results from all five auctions', *Emissions Reduction Fund*, http://www.cleanenergyregulator.gov.au/ERF/Auctions-results/april-2017.

29. Ibid.
30. Nordhaus, William (1991), 'To Slow or Not to Slow: The Economics of the Greenhouse Effect', *Economic Journal* 101, 920–37, 934, 936; Stern, Nicholas, (2006), *The Economics of Climate Change: The Stern Review* (Cambridge University Press). The starting price of carbon units under Australia's former 'carbon tax' was AUD$23 (2012), see Clean Energy Act 2011 (Cth) (repealed) s 100.
31. Clean Energy Regulator (2017), 'Emissions Reduction Fund project register: Interactive Project Map', *Emissions Reduction Fund*, http://www.cleanenergyregulator.gov.au/maps/Pages/erf-projects/index.html.
32. Australian Government (December 2016), 'Australia's Emissions Projections 2016', 25.
33. For example, the freehold owner, a tenant, or contractual licensee, see above n 18.
34. For example, scholars argue that substantial changes in the agriculture sector will not occur until the price of carbon credits exceeds $65 per tonne: Bryan, Brett A, Crossman, Neville D, Nolan, Martin, Li, Jing, Navarro, Javier and Connor, Jeffery D (2015), 'Land Use Efficiency: Anticipating Future Demand for Land-Sector Greenhouse Gas Emissions Abatement and Managing Trade-Offs with Agriculture, Water, and Biodiversity, *Global Change Biology* 21, 4098.
35. Clean Energy Regulator (2017), 'Emissions Reduction Fund project register', *Emissions Reduction Fund*, http://www.cleanenergyregulator.gov.au/ERF/project-and-contracts-registers/project-register, accessed 27 June 2017 ('ERF Register'). Note, projects are concentrated in New South Wales and Queensland.
36. Commonwealth of Australia (2016), 'Quarterly update of Australia's National Greenhouse Gas Inventory: June 2016' *Greenhouse Gas Measurement and Reporting* [Data sources: Figure 18] http://www.environment.gov.au/climate-change/greenhouse-gas-measurement/publications/quarterly-update-australias-national-greenhouse-gas-inventory-jun-2016.
37. Ibid. The Carbon Credits Act entered into force on 1 July 2012; LULUCF activities were also encouraged under the former version of the Carbon Credits Act (Carbon Farming Initiative) associated with the 'carbon tax' until 30 June 2014.
38. Department of Agriculture and Water Resources (4 May 2015), *Australia's Forests*, http://www.agriculture.gov.au/forestry/australias-forests.
39. ERF Register, above n 35.
40. Above n 36.
41. Australian Government (24 November 2015), 'Carbon Neutral Program Guidelines: Volume 4'; a list of certified business can be viewed at http://www.environment.gov.au/climate-change/carbon-neutral/carbon-neutral-program/certified-businesses; see also Intrepid Group (2016), 'Communication on progress report 2016', 19–20.
42. Australian Government, above n 41.
43. Financial Services Council (2015), 'ESG Reporting Guidelines for Australian Companies'; see also Hutley, Noel, QC and Hartford-Davis, Sebastian (2016), 'Climate change and director's duties' (Centre for Policy Development and the Future Business Council, Memorandum of Legal Opinion). For example, see disclosures and reports made by QANTAS: http://investor.qantas.com/sustainability/.
44. Corporations Act 2001 (Cth) s 763A; see further http://asic.gov.au/regulatory-resources/financial-services/carbon-markets/; http://www.cleanenergyregulator.gov.au/ERF/Want-to-participate-in-the-Emissions-Reduction-Fund/Step-4-Delivery-and-payment/ACCUs-as-financial-products.
45. Australian Securities and Investments Commission (May 2015), 'Do I need an AFS to participate in carbon markets?' (Regulatory Guide 236).
46. GreenFleet website: http://www.greenfleet.com.au/About-us/About-Greenfleet.
47. ERF Register, above n 35.
48. Greenfleet (2017), 'Partner with Greenfleet to revegetate your land', *Greenfleet*, http://www.greenfleet.com.au/Landowners.
49. auscarbon Group (2017), 'About us', *auscarbon Group*, http://www.auscarbongroup.com.au/pages/auscarbon.php.
50. ERF Register, above n 35, namely Bowgada Biodiversity Project (EOP100644); Pine

Ridge Biodiversity Project (EOP100646); Hillview Biodiversity Project (ERF101383); Terra Grata Biodiversity Project (ERF101385); Tomora Biodiversity Project (ERF101389). Note, each of these projects is subject to permanence periods of 100 years as discussed in Section 5.

51. CarbonNeutral (2017), 'Testimonials', *About Us*, https://carbonneutral.com.au/testimonials/.
52. CarbonNeutral (2017), 'Case Studies', *About Us*, https://carbonneutral.com.au/case-studies/.
53. Arguably as the latest form of entities acquiring land to generate financial returns based on forestry activities. Historically, entities have participated in forestry and agribusiness managed investment schemes to minimise tax liability. For example see Senate Economics Reference Committee (March 2016), 'Agribusiness managed investment schemes: A bitter harvest'.
54. Carbon Credits Act ss 27(4), 27(5), 54.
55. Carbon Credits Act ss 28A, 43–45A.
56. Carbon Credits Act s 97(6).
57. Carbon Credits Act s 40; for discussion of the potential impact of noting a CMO against land on the central land register see Johnston, Vanessa (2016), 'Rights and Interests in Land Created under the Carbon Credits Act 2011 (Cth): Theoretical and Practical Issues', *Australian Property Law Journal* 25, 199–219.
58. Carbon Credits Act s 97(14).
59. Carbon Credits Act ss 27(3)(e), (f). Australian Government (2014), *Emissions Reduction Fund White Paper* [5.2.3]. This is also relevant to carbon rights arising from sequestration projects under State and Territory laws, as discussed by Durrant, Nicola (2011), 'Legal Issues in Carbon Farming: Biosequestration, Carbon Pricing, and Carbon Rights', *Climate Law* 2, 515–33, 515.
60. O'Connor, Pamela, Christensen, Sharon, Duncan, WD, Phillips, Angela (2013), 'From Rights to Responsibilities: Reconceptualising Carbon Sequestration Rights in Australia', *Environmental and Planning Law Journal* 30, 403–21, 407–8; Australian Government, above n 59 [3.3.2].
61. ERF Register, above n 35.
62. Ibid: New South Wales (202), followed by Queensland (124), Western Australia (31), South Australia (21), Victoria (21), Tasmania (8), and the Australian Capital Territory (1).
63. Ibid.
64. Carbon Credits Act ss 97(2)(b), (4).
65. Carbon Credits Act s 97(2).
66. Carbon Credits Act s 97(4)(a)–(e).
67. Carbon Credits Act ss 97(9), (10).
68. Carbon Credits Act s 100.
69. *King v David Allen & Sons Billposting Ltd* [1916] 2 AC 54.
70. Pursuant to a quasi-proprietary 'mere equity' enforceable at the discretion of the courts, see further Johnston, above n 57; Carbon Credits Act s 39. Note the overlap between notation of carbon sequestration projects on land registers and the registration of legal and equitable interests in land arising from 'forest carbon' (and similar) pursuant to State and Territory legislation – see further O'Connor et al, above n 60; Hepburn, Samantha (2009), 'Carbon Rights as New Property: The Benefits of Statutory Verification', *Sydney Law Review* 31, 239.
71. Christensen, Sharon, Duncan, WD, Phillips, Angela, O'Connor, Pamela (2013), 'Issues in Negotiating a Carbon Sequestration Agreement for a Biosequestration Offsets Project', *Australian Property Law Journal* 21, 195–226, 198–9; O'Connor et al, above n 60, 414–15; Discussion of State and Territory forest carbon schemes is outside the scope of this chapter. See comprehensive discussion in Christensen et al, and O'Connor et al cited above; see also Hepburn above n 70; Durrant above n 59.
72. See further Christensen et al, above n 71.

# PART II

# Complementary tax approaches

# 6. Vehicle taxation in EU Member States

**Claudia Kettner and Daniela Kletzan-Slamanig**

## 1 INTRODUCTION

In the Paris Agreement[1] the global community agreed to hold the increase in global average temperature well below 2°C above pre-industrial levels and pursue efforts to limit the temperature increase to 1.5°C. In order to achieve these ambitious climate targets, a deep decarbonisation of our societies and economies is required. Fundamental transformations of all major emitting sectors need to begin immediately and scale up quickly. In order to reach the 2°C target, greenhouse gas emissions need to be cut by 40 to 70 per cent compared to current levels by 2050. For the 1.5°C target, zero carbon dioxide ($CO_2$) emissions need to be achieved by the middle of this century, and zero greenhouse gas emissions roughly in the 2060s.[2]

Yet, global greenhouse gas emissions are on the rise. The years 2015 to 2017 were the hottest years since the beginning of records in 1880.[3] Energy- and process-related $CO_2$ emissions account for 64 per cent of global and roughly three quarters of EU greenhouse gas emissions respectively. While energy-related greenhouse gas emissions overall are declining in the EU (by 21 per cent between 1990 and 2015), some sectors have exhibited increasing emission levels since 1990. Most notably, emissions from the transport sector rose by 17 per cent from 758 Mt $CO_2$ in 1990 to 887 Mt $CO_2$ in 2015 and they now account for more than 20 per cent of the EU's greenhouse gas emissions.[4] While energy consumption of petrol-driven cars in the EU Member States has declined since the late 1990s, diesel consumption of passenger cars and trucks in 2015 was significantly higher than in 1990.[5]

In order to reduce emissions in the transport sector, a comprehensive portfolio of policy instruments is necessary, as stressed, for instance, in the European Commission Strategy for Low Emission Mobility.[6] This includes measures ranging from efficient spatial planning via digital mobility solutions and the promotion of multi-modality to alternative drives and the provision of related infrastructure. One central lever in this context is the development of fiscal measures for the transport sector. This includes inter alia a more stringent taxation of fossil fuels or the specific emissions of vehicles as well as the introduction of other mechanisms such as road pricing to tackle other negative effects related to transport such as noise, soil sealing and congestion.

According to economic theory the optimal solution would be the taxation of $CO_2$ emissions related to energy consumption, as thereby the negative environmental effects resulting from vehicle usage are taxed, instead of mere vehicle ownership. Recent analyses[7] have shown, however, that registration taxes can contribute to reducing carbon emissions from the transport sector: since consumers tend to consider fuel cost savings for a period of only about three years in their purchase decisions[8] (instead of the expected usage costs over the whole lifetime of vehicles), upfront taxes can help counterbalancing consumer myopia. So far, only a limited number of empirical studies analysing the effect of registration and ownership taxes on $CO_2$ emissions in the transport sector is available.[9] Generally, these studies confirm the positive effect of $CO_2$-based vehicle taxes (most notably registration taxes) on emissions, albeit with variations in the magnitude of effects as they depend on the structure and design of the tax.

The remainder of this chapter is structured as follows: In Section 2, the methodological approach for calculating vehicle tax rates and implicit $CO_2$ vehicle tax rates in the EU Member States is laid out. Then purchase and ownership taxes are calculated for three classes of petrol and diesel cars and differences between the Member States are elaborated (Section 3). The final section (Section 4) summarises the main findings and concludes the chapter.

## 2 METHODOLOGICAL APPROACH AND DATA SOURCES

This chapter presents an overview of acquisition and ownership tax levels in the 28 EU Member States. Acquisition and ownership tax rates are calculated for three categories of petrol and diesel cars in order to illustrate the potential sensitivity of tax rates in the different Member States with respect to $CO_2$ emissions:

1. an average newly registered car in the Member States in terms of specific $CO_2$ emissions;

2. an average newly registered car from the first tertile in terms of specific $CO_2$ emissions; and
3. an average newly registered car from the third tertile in terms of specific $CO_2$ emissions.

In addition to the explicit tax rates per vehicle, implicit carbon tax rates of vehicles are calculated: non-recurring tax rates are divided by the assumed $CO_2$ emissions over the whole service life of the cars; recurring annual ownership tax rates are divided by the assumed annual $CO_2$ emissions of the cars (details below). Thereby we extend the OECD's concept of effective carbon rates. The OECD uses this approach in order to assess and compare the stringency of different countries' climate policy. They define effective carbon rates as the 'total price that applies to $CO_2$ emissions from energy use as a result of market-based policy instruments'.[10] This concept hence includes three categories of carbon prices, namely explicit carbon taxes, specific taxes on energy use[11] and the price of tradable emission permits.

Data on specific $CO_2$ emissions as well as on other parameters determining the tax level such as vehicle mass, power, and capacity on Member State level are taken from the European Environmental Agency's vehicle registration statistics.[12] Tables 6.1 and 6.2 summarise these characteristics of the six categories of cars in the 28 EU Member States. In addition, a service life of seven years is assumed for all cars and a travelled distance of 9,535 km p.a. for petrol and 17,245 km p.a. for diesel cars respectively, which are the average European distances of the two car types according to the Odyssee database.[13]

With respect to passenger car prices, no comprehensive database is available. Therefore in the calculations the price of an average Austrian car is used, which amounts to EUR 22,600 for diesel- and to EUR 13,200 for petrol-driven cars.[14] To account for potential variation in price levels, a sensitivity analysis is performed.

## 3 VEHICLE TAXATION IN EU MEMBER STATES

In 2016, 14 EU Member States considered the $CO_2$ emissions of a vehicle for determining registration taxes and recurring ownership taxes (Figure 6.1). In four of these countries (Cyprus, Estonia, France and Latvia), specific $CO_2$ emissions are the only criterion for acquisition taxes; in the other countries $CO_2$ emissions are combined with other factors (such as purchase prices or cylinder capacity). In Cyprus, France, Croatia and Ireland, $CO_2$ emissions are the sole criterion for determining the level of

*Table 6.1 Characteristics of petrol-driven cars in EU Member States*

| | Mass | | | Capacity | | | Power | | |
|---|---|---|---|---|---|---|---|---|---|
| | Average | Tertile 1 | Tertile 3 | Average | Tertile 1 | Tertile 3 | Average | Tertile 1 | Tertile 3 |
| AT | 1,261 | 1,069 | 1,474 | 1,323 | 1,072 | 1,722 | 77 | 74 | 148 |
| BE | 1,245 | 1,035 | 1,470 | 1,337 | 1,063 | 1,714 | 81 | 73 | 128 |
| BG | 1,217 | 1,051 | 1,514 | 1,422 | 1,077 | 2,045 | 87 | 69 | 162 |
| CY | 1,248 | 1,065 | 1,448 | 1,391 | 1,120 | 1,746 | 85 | 58 | 120 |
| CZ | 1,225 | 1,066 | 1,407 | 1,348 | 1,141 | 1,741 | 81 | 74 | 153 |
| DE | 1,275 | 1,028 | 1,529 | 1,463 | 1,076 | 1,997 | 95 | 83 | 160 |
| DK | 1,123 | 944 | 1,349 | 1,206 | 988 | 1,524 | 69 | 72 | 137 |
| EE | 1,365 | 1,169 | 1,576 | 1,589 | 1,181 | 2,061 | 99 | 76 | 154 |
| ES | 1,204 | 1,020 | 1,425 | 1,308 | 1,076 | 1,646 | 80 | 56 | 113 |
| FI | 1,306 | 1,110 | 1,492 | 1,406 | 1,104 | 1,786 | 88 | 82 | 145 |
| FR | 1,159 | 991 | 1,357 | 1,239 | 1,037 | 1,546 | 75 | 65 | 119 |
| GR | 1,138 | 1,009 | 1,386 | 1,217 | 1,058 | 1,505 | 69 | 55 | 96 |
| HR | 1,166 | 954 | 1,337 | 1,287 | 996 | 1,590 | 70 | 60 | 117 |
| HU | 1,251 | 1,012 | 1,438 | 1,461 | 1,067 | 1,789 | 87 | 66 | 149 |
| IE | 1,156 | 1,074 | 1,339 | 1,241 | 1,177 | 1,509 | 76 | 76 | 104 |
| IT | 1,097 | 1,028 | 1,267 | 1,230 | 1,080 | 1,541 | 62 | 56 | 120 |
| LT | 1,262 | 970 | 1,546 | 1,504 | 1,014 | 1,973 | 92 | 61 | 136 |
| LU | 1,321 | 976 | 1,655 | 1,676 | 1,204 | 2,593 | 121 | 63 | 159 |
| LV | 1,350 | 1,042 | 1,583 | 1,572 | 1,098 | 2,113 | 96 | 76 | 161 |
| MT | 1,070 | 1,148 | 1,272 | 1,211 | 1,193 | 1,447 | 74 | 76 | 100 |
| NL | 1,267 | 930 | 1,432 | 1,334 | 995 | 1,654 | 75 | 52 | 124 |
| PL | 1,290 | 973 | 1,511 | 1,468 | 977 | 1,898 | 89 | 72 | 149 |
| PT | 1,129 | 1,076 | 1,278 | 1,156 | 1,133 | 1,461 | 66 | 72 | 124 |
| RO | 1,174 | 992 | 1,371 | 1,278 | 966 | 1,631 | 73 | 67 | 136 |
| SE | 1,323 | 1,031 | 1,540 | 1,471 | 1,071 | 1,984 | 92 | 61 | 150 |
| SI | 1,192 | 1,090 | 1,368 | 1,257 | 1,110 | 1,525 | 75 | 94 | 125 |
| SK | 1,265 | 1,051 | 1,444 | 1,375 | 1,086 | 1,732 | 80 | 74 | 152 |
| UK | 1,220 | 1,112 | 1,466 | 1,396 | 1,143 | 1,843 | 85 | 74 | 127 |
| **EU** | **1,223** | **1,016** | **1,446** | **1,359** | **1,060** | **1,794** | **83** | **55** | **123** |

*Source:* European Environmental Agency (2016), *Monitoring of $CO_2$ Emissions from Passenger Cars*, Data 2015.

recurring ownership taxes.[15] In five countries, no other taxes than value added tax are levied on the purchase of new vehicles (Bulgaria, Germany, Luxembourg, Sweden and the UK), likewise two Member States do not collect recurring vehicle ownership taxes (Poland and Slovakia). In three Member States, neither registration taxes nor ownership taxes are implemented as of 2016 (the Czech Republic, Estonia and Lithuania).

*Table 6.2 Characteristics of diesel-driven cars in EU Member States*

| | Mass | | | Capacity | | | Power | | |
|---|---|---|---|---|---|---|---|---|---|
| | Average | Tertile 1 | Tertile 3 | Average | Tertile 1 | Tertile 3 | Average | Tertile 1 | Tertile 3 |
| AT | 1,640 | 1,394 | 1,932 | 1,837 | 1,536 | 2,234 | 100 | 53 | 110 |
| BE | 1,532 | 1,304 | 1,781 | 1,746 | 1,486 | 2,156 | 93 | 54 | 116 |
| BG | 1,543 | 1,242 | 1,914 | 1,839 | 1,470 | 2,426 | 103 | 59 | 138 |
| CY | 1,602 | 1,340 | 1,888 | 1,845 | 1,530 | 2,280 | 103 | 71 | 140 |
| CZ | 1,562 | 1,306 | 1,879 | 1,868 | 1,540 | 2,376 | 105 | 60 | 110 |
| DE | 1,633 | 1,373 | 1,943 | 1,995 | 1,768 | 2,476 | 118 | 57 | 148 |
| DK | 1,453 | 1,280 | 1,699 | 1,728 | 1,481 | 2,108 | 94 | 49 | 98 |
| EE | 1,636 | 1,337 | 2,011 | 1,875 | 1,496 | 2,411 | 106 | 72 | 133 |
| ES | 1,448 | 1,227 | 1,686 | 1,711 | 1,481 | 2,080 | 95 | 68 | 128 |
| FI | 1,635 | 1,428 | 1,865 | 1,908 | 1,560 | 2,220 | 112 | 63 | 116 |
| FR | 1,420 | 1,200 | 1,669 | 1,669 | 1,447 | 2,078 | 88 | 54 | 111 |
| GR | 1,330 | 1,156 | 1,535 | 1,482 | 1,312 | 1,687 | 75 | 50 | 99 |
| HR | 1,427 | 1,237 | 1,654 | 1,693 | 1,476 | 2,041 | 87 | 53 | 97 |
| HU | 1,595 | 1,316 | 1,927 | 1,873 | 1,540 | 2,282 | 106 | 64 | 114 |
| IE | 1,480 | 1,291 | 1,706 | 1,702 | 1,470 | 2,050 | 95 | 68 | 126 |
| IT | 1,438 | 1,199 | 1,713 | 1,650 | 1,369 | 2,063 | 86 | 50 | 85 |
| LT | 1,587 | 1,320 | 1,942 | 1,891 | 1,523 | 2,450 | 105 | 70 | 151 |
| LU | 1,565 | 1,293 | 1,879 | 1,925 | 1,517 | 2,460 | 113 | 61 | 209 |
| LV | 1,649 | 1,347 | 2,044 | 1,953 | 1,524 | 2,522 | 110 | 69 | 132 |
| MT | 1,402 | 1,149 | 1,739 | 1,694 | 1,440 | 2,102 | 94 | 67 | 129 |
| NL | 1,443 | 1,273 | 1,652 | 1,731 | 1,509 | 2,061 | 89 | 52 | 112 |
| PL | 1,584 | 1,320 | 1,903 | 1,849 | 1,522 | 2,254 | 105 | 60 | 125 |
| PT | 1,438 | 1,245 | 1,671 | 1,640 | 1,416 | 2,076 | 88 | 51 | 99 |
| RO | 1,464 | 1,199 | 1,806 | 1,747 | 1,443 | 2,341 | 93 | 53 | 101 |
| SE | 1,673 | 1,460 | 1,883 | 2,016 | 1,760 | 2,381 | 122 | 63 | 132 |
| SI | 1,476 | 1,266 | 1,721 | 1,735 | 1,480 | 2,226 | 94 | 61 | 99 |
| SK | 1,620 | 1,340 | 1,955 | 1,875 | 1,545 | 2,360 | 105 | 60 | 106 |
| UK | 1,573 | 1,314 | 1,868 | 1,899 | 1,530 | 2,371 | 106 | 71 | 146 |
| **EU** | **1,525** | **1,287** | **1,801** | **1,804** | **1,571** | **2,241** | **100** | **73** | **138** |

*Source:* European Environmental Agency (2016), *Monitoring of $CO_2$ Emissions from Passenger Cars*, Data 2015.

## 3.1 Taxation of Petrol-driven Cars

### 3.1.1 Acquisition and ownership tax rates of petrol-driven cars

Not only the structure of vehicle taxation differs considerably between EU Member States, but also the tax rates. For an average petrol-driven car from the EU with 83 kW and a capacity of 1,360 $cm^3$ emitting 122 grams of $CO_2$ per kilometre, registration taxes range from EUR 0 in France to EUR 22,000 in Denmark. In France a bonus-malus system for registration taxes was introduced in 2008, which means that for passenger cars registered in

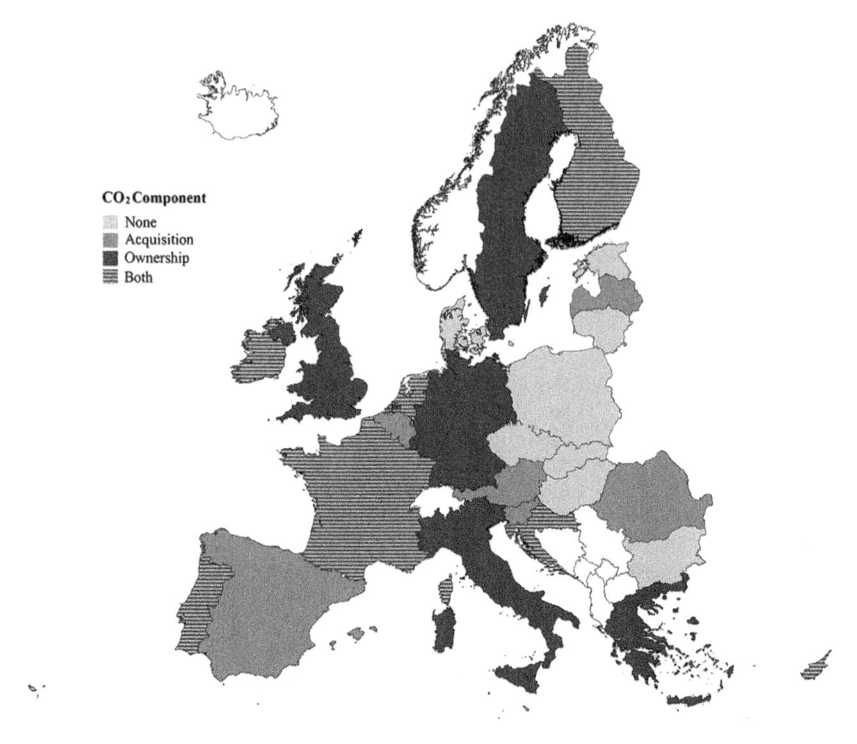


*Source:* Own illustration based on ACEA (2016), *ACEA Tax Guide 2016*.

*Figure 6.1 $CO_2$ components in vehicle taxation in EU Member States*

France the tax depends on the vehicle $CO_2$ emissions. In 2016, no tax was due for vehicles with specific $CO_2$ emissions of no more than 130 grams per kilometre.[16] As of 2016, there are a total of 12 different tax rates reflecting different categories of emission intensity. In the highest tax category with emissions of more than 250 grams per kilometre, the French registration tax amounts to EUR 8,000. In Spain vehicle registration tax rates are also based on specific $CO_2$ emissions with cars emitting less than 120 grams of $CO_2$ per kilometre being exempt from the tax. For more emission-intensive passenger cars the tax rates increase to up to 14.75 per cent. The Danish vehicle registration tax in contrast is based on the purchase price. For cars with a taxable value[17] of below DKK 81,700 (EUR 10,970) the tax rate applied is 105 per cent; for more expensive cars the tax rate is 105 per cent of DKK 81,700 (EUR 10,970) plus 150 per cent of the remaining taxable value. The taxable vehicle value and hence the registration tax are reduced or increased based on a number of factors, related to the safety equipment and evaluation of the cars as well as to fuel efficiency.[18] Moreover, a minimum tax rate of DKK 20,000 (EUR 2,686) for all private passenger cars is in place. In addition to Denmark, price is an important factor in determining the registration tax rate in eight other EU Member States, often also in combination with specific $CO_2$ emissions such as in Austria, Croatia, Finland, Ireland or Spain. When all other characteristics of the cars are equal, variations in registration taxes in these countries are directly proportional to variations in car prices.

Also with respect to recurring ownership taxes, pronounced differences between Member States can be identified. As of 2016 vehicle ownership tax rates are in the range from EUR 0 to EUR 600 per annum for the average EU petrol car. The lowest recurring taxes are levied in France and the UK, while the highest ownership taxes are found in the Netherlands. In France annual ownership taxes of EUR 160 only have to be paid for passenger cars whose specific emissions exceed 190 grams $CO_2$ per kilometre since 2012. In the UK annual ownership tax rates are scaled depending on the specific $CO_2$ emissions of the cars: up to specific emissions of 100 grams $CO_2$ per kilometre, no ownership taxes need to be paid; if specific emissions exceed 255 grams per kilometre the tax rate is GBP 505 (EUR 617) per annum.

### 3.1.2 Implicit $CO_2$ taxation of petrol-driven vehicles

Implicit $CO_2$ acquisition and vehicle tax rates basically reflect three underlying factors: (1) the overall registration or ownership tax rates per vehicle, (2) the assumed distance travelled with the vehicle over a certain period of time (or more specifically over the whole service life of the vehicle in the case of registration taxes and annually in the case of ownership taxes), and (3) the specific $CO_2$ emissions of the vehicle. Currently tax rates are often linked to the car's specific $CO_2$ emissions, either directly with specific $CO_2$ emissions constituting a tax base component or indirectly via the vehicles' power or capacity, which are correlated with emission intensity.

Figure 6.2 shows implicit $CO_2$ tax rates payable at the registration of vehicles. The figure details the three categories of vehicles specific to the Member States as well as respective tax rates for the average EU petrol car. Again, pronounced differences between Member States can be noted with implicit purchase-related carbon taxes between EUR 0 and EUR 3,060 per tonne $CO_2$.[19] Moreover, a significant spread in implicit $CO_2$ registration taxes can be observed within some Member States, especially in Cyprus, Denmark and the Netherlands. While in Cyprus and in the Netherlands the most emission-intensive cars, namely cars from the third tertile, bear the largest tax burden, since the tax rate increases at a higher rate than the cars' specific $CO_2$ emissions, the opposite is true

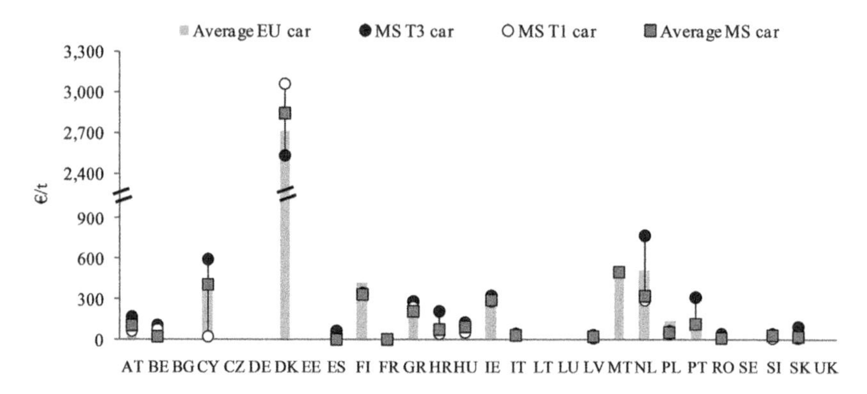


*Source:* Own calculations based on European Environmental Agency (2016), *Monitoring of $CO_2$ Emissions from Passenger Cars*, Data 2015; ACEA (2016), *ACEA Tax Guide 2016*.

*Figure 6.2 Implicit purchase $CO_2$ tax rates of petrol-driven cars*

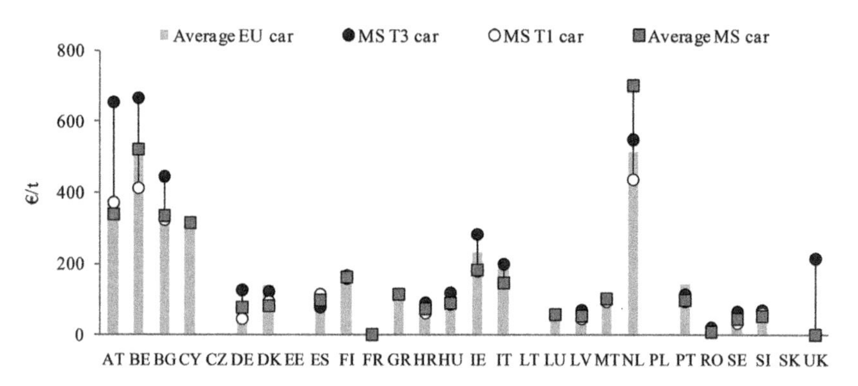


*Source:* Own calculations based on European Environmental Agency (2016), *Monitoring of $CO_2$ Emissions from Passenger Cars*, Data 2015; ACEA (2016), *ACEA Tax Guide 2016*.

*Figure 6.3 Implicit ownership $CO_2$ tax rates of petrol-driven cars*

for Denmark, where registration tax rates depend only on the vehicles' purchase prices.

In Figure 6.3 ownership tax rates per tonne $CO_2$ emitted are displayed. The highest implicit ownership tax rates for petrol-driven cars are added in the Netherlands, Belgium and Austria and also the spread in tax rates is considerable within these countries. For the Netherlands it is worth noting that by far the highest ownership tax rates per tonne $CO_2$ have to be paid for the average Dutch petrol car. The Netherlands, moreover, impose

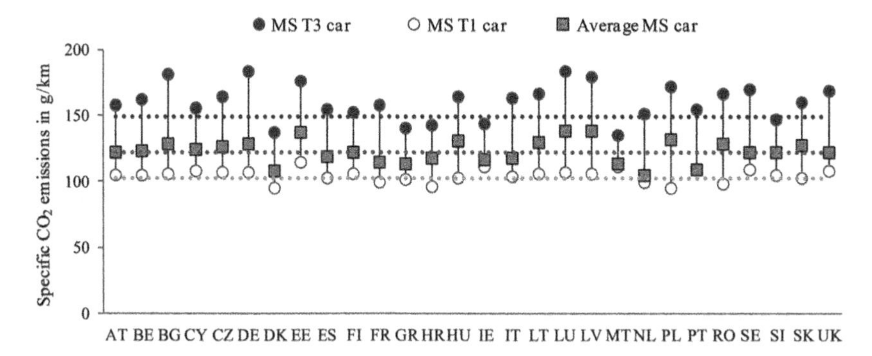


*Source:* Own calculations based on European Environmental Agency (2016), *Monitoring of $CO_2$ Emissions from Passenger Cars*, Data 2015; ACEA (2016), *ACEA Tax Guide 2016*.

*Figure 6.4 Specific $CO_2$ emissions of petrol-driven cars*

the highest tax rates on petrol consumption in the EU.[20] A considerable spread in tax rates is also found for the UK, where no ownership taxes have to be paid for the low- and average-emission-intensive car, while for the car from the most emission-intensive tertile annual ownership taxes amount to EUR 215.

### 3.1.3 Specific $CO_2$ emissions of petrol-driven vehicles

The specific $CO_2$ emissions of new petrol-driven cars are shown in Figure 6.4. Comparably low emission intensities can be found in Denmark, France, Greece, Croatia, Ireland, the Netherlands and Portugal. On average, the specific emissions are lowest in Denmark, the Netherlands and Portugal. Malta, Denmark and Greece show the lowest specific emissions from the most emission-intensive tertile.

## 3.2 Taxation of Diesel-driven Cars

### 3.2.1 Acquisition and ownership tax rates of diesel-driven cars

In the EU, the average diesel-driven car has a higher mass, capacity and power than the average petrol car; in terms of the specific $CO_2$ emissions the performance of the average diesel car is, however, better. The average diesel-driven car weighs 1,300 kilograms and has 100 kW and a capacity of 1,800 $cm^3$; the specific emissions are 119 grams of $CO_2$ per kilometre.

For an average diesel car in the EU, registration taxes are zero in France and Spain and highest in Denmark with EUR 38,500. That no registration taxes for the average diesel-driven car have to be paid in France and Spain is again the result of the tax exemptions for cars emitting less than 130 and 120 grams per kilometre respectively.

Recurring ownership taxes also show pronounced differences between Member States; in 2016 vehicle ownership tax rates ranged between EUR 0 and EUR 1,960 for the average diesel car. As for petrol-driven cars the lowest recurring taxes are levied in France and the UK (where emission-efficient cars emitting less than a certain amount of $CO_2$ per kilometre are exempt from paying ownership taxes), while the highest ownership taxes are found in the Netherlands. Notably, ownership tax rates in the Netherlands are significantly higher for diesel-driven cars than for petrol-driven cars.

### 3.2.2 Implicit $CO_2$ taxation of diesel-driven vehicles

$CO_2$ tax rates payable at the registration of diesel cars are displayed in Figure 6.5 for the three categories of cars in the Member States as well as the respective tax rates for the average EU diesel car. Again, pronounced differences between Member States can be noted with implicit purchase carbon taxes between EUR 0 and EUR 3,100 per tonne $CO_2$. Significant spreads of implicit $CO_2$ registration taxes can be observed in Cyprus, Denmark and the Netherlands.

Figure 6.6 shows ownership tax rates per tonne $CO_2$ emitted in the Member States by diesel cars. By far the highest implicit ownership tax rates for diesel-driven cars have to be paid in the Netherlands (with EUR 910 for an average car and EUR 980 for an emission-intensive diesel

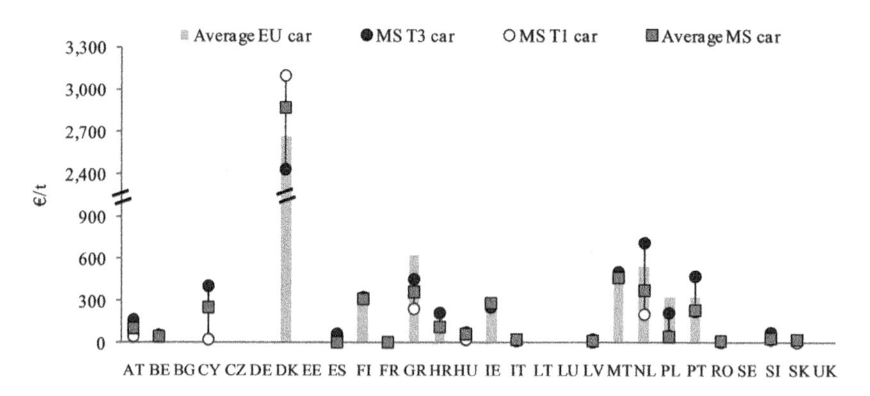


*Source:* Own calculations based on European Environmental Agency (2016), *Monitoring of $CO_2$ Emissions from Passenger Cars*, Data 2015; ACEA (2016), *ACEA Tax Guide 2016*.

*Figure 6.5 Implicit purchase $CO_2$ tax rates of diesel-driven cars*

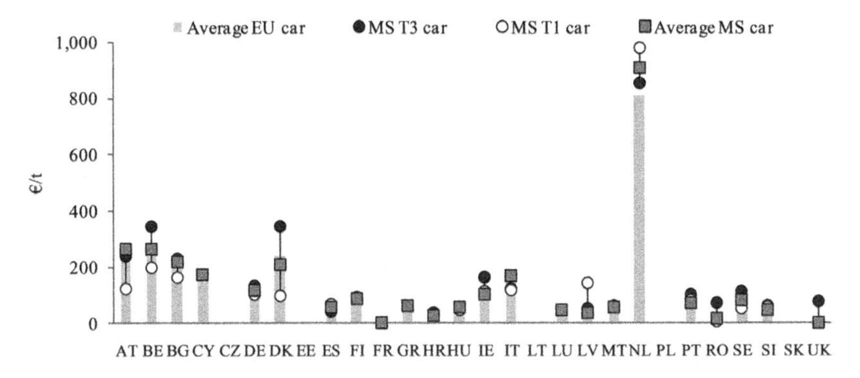


*Source:* Own calculations based on European Environmental Agency (2016), *Monitoring of $CO_2$ Emissions from Passenger Cars*, Data 2015; ACEA (2016), *ACEA Tax Guide 2016*.

*Figure 6.6 Implicit ownership $CO_2$ tax rates of diesel-driven cars*

car), followed by Belgium, Austria and Denmark. The highest spread in ownership tax rates is found for Denmark, where ownership tax rates are determined on the basis of fuel consumption and vehicle mass. Implicit $CO_2$ car ownership tax rates for diesel cars in Denmark range from EUR 100 per tonne $CO_2$ for the least emission-intensive car to EUR 350 per tonne $CO_2$ for the most emission-intensive car.

### 3.2.3 Specific $CO_2$ emissions of diesel-driven vehicles

Figure 6.7 shows the specific $CO_2$ emissions of newly registered diesel cars. By far the lowest emission intensities are found for the Netherlands, with 81 grams $CO_2$ per kilometre for the low-emission car, 133 for the emission-intensive car and 97 for the average diesel car. This supports the general finding in the literature that registration and ownership taxes can influence purchase decisions since the Netherlands levy the highest ownership taxes and the second highest registration taxes of all EU Member States, but at the same time the lowest diesel tax rates. The Netherlands deliberately introduced higher registration and ownership taxes for diesel cars in order to compensate for the low taxes on diesel consumption. Low specific emissions for the average and the most emission-intensive diesel car are also found for Croatia, Denmark, France, Greece and Portugal.

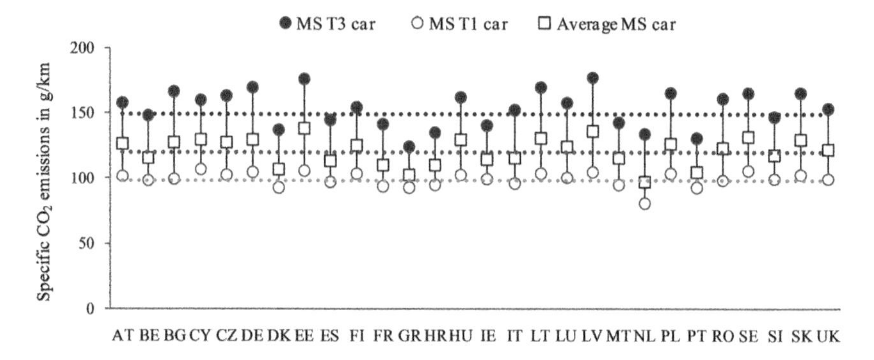


*Source:* Own calculations based on European Environmental Agency (2016), *Monitoring of $CO_2$ Emissions from Passenger Cars*, Data 2015; ACEA (2016), *ACEA Tax Guide 2016*.

*Figure 6.7 Specific $CO_2$ emissions of diesel-driven cars*

# 4 SUMMARY AND CONCLUDING REMARKS

In order to mitigate the detrimental effects of global warming, a comprehensive transformation of our societies is inevitable. In all major emitting sectors, greenhouse gas emissions have to be reduced drastically by the middle of the century; in this respect energy-producing and -consuming sectors play an important role since they are a dominant source of greenhouse gas emissions worldwide.

The transport sector is of particular relevance: accounting for 24 per cent of global energy-related $CO_2$ emissions (or roughly one-sixth of total greenhouse gas emissions) and growing continuously, the sector requires a comprehensive bundle of policy measures in order to achieve a decarbonisation. This includes both the introduction of emission standards for vehicles, infrastructural approaches, strategies regarding the phasing out of fossil fuel-based vehicles, and respectively a prohibition of fossil fuel-based vehicles as well as the introduction of an adequate tax structure to incentivise behavioural changes in the sector.

According to economic theory the first best solution would be the introduction of a carbon tax levied directly on the $CO_2$ emissions, and respectively on the consumption of fossil fuels, since fossil fuel use is related to $CO_2$ emissions in direct proportion. Compared to an emission standard, taxes are characterised by a higher degree of flexibility, implying that a given emission target can be achieved at a lower cost.[21]

In the EU, explicit carbon taxes on fossil fuel use have only been implemented in ten Member States. These carbon taxes, however, only account for a small portion of the total excise duties on fossil fuels in the transport sector, with the exception of Sweden, where the carbon tax accounts for approximately half of the total excise duties on petrol and diesel. Implicit carbon tax rates on transport fuels – energy tax rates converted into carbon tax rates via energy content and emission intensity – currently do not reflect differences in the emission intensity of fuels. With respect to the transport sector diesel is taxed at a (sometimes considerably) lower rate in all EU Member States due to competitiveness concerns and respectively preferential tax rates for the diesel-consuming freight transport sector.

Against this background, vehicle taxes based on $CO_2$ emissions can be a useful complement for reducing emissions from passenger cars. Specific $CO_2$ emissions are considered in vehicle taxation in 20 Member States, with pronounced differences in tax levels and spreads. Our descriptive analysis suggests that both the level of registration and ownership tax rates show a positive correlation with the specific $CO_2$ emissions of newly registered vehicles. One noteworthy example in this context is the Netherlands, which levies the highest ownership taxes and the second highest registration taxes of all EU Member States despite the lowest excise duties on diesel consumption. While vehicle taxation in the Netherlands is inter alia based on cars' $CO_2$ emissions, evidence from other EU Member States shows that the mere magnitude of vehicle tax rates, as in Denmark, is also positively correlated with low $CO_2$ intensities. Overall cars' $CO_2$ intensities tend to be lower in Member States with higher ownership and especially purchase taxes but high vehicle tax levels alone do not guarantee a shift towards low-emission vehicles.

When implementing $CO_2$-based vehicle taxes, the design of the tax deserves careful consideration in order to avoid unintended side-effects: calculating tax rates merely on the basis of $CO_2$ emissions might imply an increase in the share of diesel-driven cars, which goes hand in hand with increasing $NO_x$ and particulate matter emissions.[22]

## ACKNOWLEDGEMENTS

The research is part of the project 'CATs – Carbon Taxes in Austria: Implementation Issues and Impacts', which was funded by the Austrian 'Klima- und Energiefonds' and carried out within the Austrian Climate Research Programme ACRP. We thank Katharina Köberl for excellent research assistance.

# NOTES

1. UNFCCC (2015), Adoption of the Paris Agreement, Paris.
2. See Kuramochi, T. et al. (2016), *The Ten Most Important Short-term Steps to Limit Warming to 1.5°C*, Cologne.
3. PBL Netherlands Environmental Assessment Agency (2016), *Trends in Global $CO_2$ Emissions: 2016 Report*, The Hague; World Meteorological Organization (2017), '2017 is set to be in top three hottest years, with record-breaking extreme weather', press release published 6 November 2017.
4. UNFCCC (2017), *National Inventory Reports 2017*.
5. Odyssee Database (2017), http://www.odyssee-mure.eu/, accessed 7 September 2017.
6. European Commission (2016), *A European Strategy for Low-Emission Mobility*, COM (2016) 501 final.
7. See e.g. Allcott, H. and N. Wozny (2014), 'Gasoline Prices, Fuel Economy, and the Energy Paradox', *Review of Economics and Statistics*, 96(5), 779–95; Brand, C., J. Anable and M. Tran (2013), 'Accelerating the Transformation to a Low Carbon Passenger Transport System: The Role of Car Purchase Taxes, Feebates, Road Taxes and Scrappage Incentives in the UK', *Transportation Research Part A: Policy and Practice*, 49, 132–48; Greene, D.L., P.D. Patterson, M. Singh and J. Li (2005), 'Feebates, Rebates and Gas-guzzler Taxes: A Study of Incentives for Increased Fuel Economy', *Energy Policy*, 33(6), 757–75; Gallagher, K.S. and E.J. Muehlegger (2011), 'Giving Green to Get Green. Incentives and Consumer Adoption of Hybrid Vehicle Technology', *Journal of Environmental Economics and Management*, 61(1), 1–15; Dineen, D., L. Ryan and B. Ó Gallachóir (2017), 'Vehicle Tax Policies and New Passenger Car $CO_2$ Performance in EU Member States', *Climate Policy*, 18(4), 369–412.
8. See e.g. Greene et al. n. 7 above; Greene, D., D.H. Evans and J. Hiestand (2013), 'Survey Evidence on the Willingness of U.S. Consumers to Pay for Automotive Fuel Economy', *Energy Policy*, 61, 1539–50.
9. For an overview see Kettner, C. and D. Kletzan-Slamanig (2017), 'Carbon Taxation in EU Member States: Evidence from the Transport Sector', in Weishaar, S.E., L. Kreiser, J.E. Milne, H. Ashiabor and M. Mehling (eds.), *The Green Market Transition. Carbon Taxes, Energy Subsidies and Smart Instrument Mixes, Critical Issues in Environmental Taxation Vol. XIX*, Northampton, MA, USA and Cheltenham, UK: Edward Elgar Publishing, pp. 17–29.
10. OECD (2016), *Effective Carbon Rates. Pricing $CO_2$ through Taxes and Emissions Trading Systems*, Paris.
11. Mainly excise taxes.
12. European Environmental Agency (2016), *Monitoring of $CO_2$ Emissions from Passenger Cars*, Data 2015.
13. Odyssee Database (2017), http://www.odyssee-mure.eu/, accessed 7 September 2017.
14. Average prices for the different vehicle categories are based on Autorevue (2017), *Alle Autos in Österreich. Der komplette Katalog – Preise/Daten/Verbrauch*, Vienna.
15. See Kettner and Kletzan-Slamanig n. 9 above, for a more detailed analysis of the structure of tax base components in EU vehicle taxation.
16. The tax exemption limit was gradually reduced from 160 grams per kilometre in 2008/2009 to 130 grams per kilometre in 2014/2016.
17. The taxable value is defined as the dealers' sales price including a profit margin of at least 9 per cent.
18. Different efficiency thresholds for diesel- and petrol-driven cars apply.
19. The lower tax rate for the emission-intensive (T3) car in Denmark reflects that the same price is assumed for all categories of vehicles, which results in the same registration tax rate. The identical tax rate in conjunction with higher specific $CO_2$ emissions translates into a lower implicit $CO_2$ registration tax rate.
20. See Kettner and Kletzan-Slamanig n. 9 above, for an overview of petrol and diesel tax rates in the EU Member States.

21. The same is true for carbon trading systems which could as upstream systems, that is at the level of fuel suppliers, be easily implemented for the road transport sector. While voluntary carbon trading is discussed e.g. for the aviation sector, for the road transport sector taxes are more frequently discussed in politics.
22. E.g. Rogan, F., E. Dennehy, H. Daly, M. Howley and B.P. Ó Gallachóir (2011), 'Impacts of an Emission Based Private Car Taxation Policy. First Year Ex-post Analysis', *Transportation Research Part A: Policy and Practice*, 45(7), 583–97; Leinert, S., H. Daly, B. Hyde and B. Ó Gallachóir (2013), 'Co-benefits? Not Always: Quantifying the Negative Effect of a $CO_2$-reducing Car Taxation Policy on $NO_x$ Emissions', *Energy Policy*, 63, 1151–9; Gerlagh, R., I. van den Bijgaart, H. Nijland and T. Michielsen (2016), 'Fiscal Policy and $CO_2$ Emissions of New Passenger Cars in the EU', *Environmental and Resource Economics*, 65(3), 1–32.

# 7. Cutting Europe's lifelines to coal subsidies

**Shelagh Whitley, Laurie van der Burgh, Leah Worrall and Sejal Patel**

## 1 BACKGROUND

### 1.1 Europe's Shift Away from Coal

With the Paris Agreement on climate change coming into force in 2016, world leaders not only reaffirmed their commitment to limit the increase in global average temperature to well below 2°C degrees, but also agreed to pursue efforts to limit global temperature rise to 1.5°C target (UNFCCC, 2015).

If countries are to meet these commitments, at least three-quarters of the existing proven reserves of oil, gas and coal need to be left in the ground, and an urgent shift to low-carbon energy is imperative (IPCC, 2014). As coal produces higher greenhouse gas (GHG) emissions when burned than oil or gas – even in a 2°C warmer world – nearly all coal resources need to remain unutilised (McGlade and Ekins, 2015).

In terms of the implications for coal-fired power, the International Energy Agency (IEA) estimates that to get to 2°C degrees, coal power plant emissions in Europe must fall by 80 per cent, and more than halve globally (a 54 per cent drop) by 2030 (IEA, 2016). A recent study by Carbon Tracker and the Grantham Institute indicates that global coal demand will peak in three years (by 2020) (Sussams and Leaton, 2017). Climate Analytics has also estimated that a full coal phase-out by 2030 will be the cheapest way for Europe to meet the 1.5°C target and that Europe should replace this capacity with renewables and energy efficiency measures (Rocha et al., 2017). Germany and Poland have the most work to do on this coal phase-out, as they are jointly responsible for 51 per cent of installed coal capacity and 54 per cent of coal emissions (Rocha et al., 2017).

A coal phase-out is also vital for helping improve air quality in Europe.

Air pollution is the single largest environmental cause of premature death in the urban parts of the continent and emissions from coal plants are partly responsible for this, with around 23,000 early deaths every year because of coal burning (Jones et al., 2016). In February 2017, the European Commission ruled that 23 European Union (EU) countries have been breaking air quality laws, through emissions from vehicles, power plants, smelting and refuse burning (Crisp, 2017).

Driven by decarbonisation objectives and policies, as well as a sharp reduction in the cost of renewable energy technologies, electricity markets across Europe and other regions around the world are going through significant transformation (van der Burg and Whitley, 2016). A recent study has found that 92 gigawatts (GW) of coal plant capacity was halted in the EU between 2010 and 2016, with only 25 GW implemented over the same time period (Shearer et al., 2017).

These changing conditions are already altering utility business models in Europe. In 2016, Germany's two large power generators E.ON and RWE both announced that they would split off their conventional power production from their businesses focused on renewables[1] (Chazan, 2016). In 2015, Italy's largest utility, Enel, agreed to phase out new investments in coal and, in 2016, Denmark's DONG Energy committed to a phase-out by 2023: a company that still had coal plants in the planning phase less than ten years ago (Clark, 2017). These trends across the power sector are likely to become more widespread throughout Europe, with several countries having already achieved a coal phase-out or committed to ending coal-fired power between 2023 and 2030 (see Box 7.1).

Despite these high-level pledges, governments have often used the energy transition, including a shift to renewables, as a justification for extending and introducing new subsidies to coal. Whether intentionally or not, these subsidies are now paying polluters and slowing the transition. The phasing out of these subsidies is widely agreed to be critical for the energy transition and to ensure financial and economic sustainability, fight air pollution and achieve climate targets. It also presents an opportunity for Europe to demonstrate leadership.

## 1.2 Europe's Fossil Fuel Subsidy Commitments

Prior to the Paris Agreement coming into force, European countries had already made repeated commitments to end fossil fuel subsidies.

- The European Commission has repeatedly called upon Member States to phase out environmentally harmful subsidies by 2020, including those for fossil fuels, and has made a commitment to

BOX 7.1 COAL PHASE-OUT COMMITMENTS ACROSS EUROPE

Many countries and regions in Europe have already ended the use of coal-fired power, including Belgium, Cyprus, Luxembourg, Malta, Scotland and the Baltic countries. A number have already announced their intention to phase out coal in the electricity sector in future decades:

- France – by no later than 2023
- The UK and Ireland – by 2025, with Austria and Denmark also likely to end coal use by around 2025
- Sweden – in the next decade
- Finland and Portugal – by 2030.

Also, the German Climate Action Plan 2050 includes a target that comes close to halving power sector emissions between 2014 and 2030.

*Sources:* DeSmogBlog, 2016; Rocha et al., 2017; Madson, 2017; Littlecott, 2017.

remove those to hard coal mining by 2018 (European Council, 2010; European Commission, 2011).

- Internationally, the EU has:
  - committed to phasing out inefficient fossil fuel subsidies by 2025 through the G7 (G7, 2017),
  - reiterated its commitment to phase out inefficient fossil fuel subsidies every year since 2009, as part of the G20 (G20, 2016).
- All EU countries have committed to the Sustainable Development Goals (SDGs), which highlight phasing out fossil fuel subsidies as a means of implementing Goal 12 to 'ensure sustainable production and consumption patterns' (UN, 2015).

Under the Europe 2020 Strategy launched in 2010, EU Member States committed to developing plans for phasing out fossil fuel subsidies by 2020, with progress to be monitored under the European Semester.[2]

However, the decision was taken to remove the focus on energy and fossil fuel subsidies from the European Semester in 2015, and no new system for governing the fossil fuel phase-out has been advanced since (Sartor and Spencer, 2015). The European Commission has also been sporadic in estimating and reporting such subsidies, with its last report released in 2014 (holding data up to 2012) and no plans to update this information (Alberici et al., 2014).

This accountability gap comes despite a 2016 Commission Report emphasising that 'fossil-fuel subsidies are particularly problematic, as they disadvantage clean energy and hamper the transition to a low-carbon economy', and that 'the recent relative fall in energy prices should make it easier for governments to remove tax exemptions and other energy demand subsidies' (European Commission, 2016).

## 1.3 Tracking Coal Subsidies in Europe

The Overseas Development Institute (ODI) sought to identify and estimate the value of ongoing subsidies to coal across ten European countries (reviewing budgetary support and tax breaks).[3] The countries reviewed include the Czech Republic, France, Germany, Greece, Hungary, Italy, the Netherlands, Poland, Spain and the United Kingdom – representing 84 per cent of Europe's energy-related GHG emissions in 2012 (World Resources Institute, 2015). (Detailed information has been compiled within ten country briefs, available on the ODI website.)

This information was collected from publicly available sources, including the OECD, European Commission, IEA and public budget documents. The country data was peer reviewed by subsidy experts from each of the countries. Here, the findings from those ten studies is summarised and recommendations are made for how countries across Europe can lead the phase-out of coal subsidies.

The coal subsidies reviewed have been further categorised in terms of their role in supporting the following:

1. coal mining
2. decommissioning and environmental rehabilitation
3. transition support (individuals and communities)
4. refining and processing
5. capacity mechanisms (see Box 7.2)
6. biomass co-firing
7. coal-fired power (other) in addition to that under categories 5 and 6
8. EU Emissions Trading Scheme (EU ETS)
9. industry
10. households
11. research and development.

### BOX 7.2 ENDING SUBSIDIES TO COAL MINING IN GERMANY

In 2010, with the EU commitment to end hard coal mining by 2018, the EU also allowed support for an uneconomic activity to continue for an additional eight years.

The largest subsidy to coal identified in this review of ten European countries provides a key example. Combined Aid in North Rhine Westphalia primarily supports the sale of coal from German hard coal mines to electricity and steel producers. This subsidy has an estimated annual cost of €1.9 billion and the German government spent an estimated €18.6 billion in total between 2005 and 2014. In addition, we understand that the German hard coal corporation RAG AG provides some level of limited matching funding.

This begs the question of whether these government resources have been used in the best possible way to support a just transition away from coal, instead of propping up an uneconomic industry. It also raises the question of what the balance of responsibilities should be in shutting down coal mines and to what extent companies should set aside resources to cover these costs. Given existing EU commitments, it will be important to draw lessons from the German experience; not only to manage the ongoing transition away from coal mining, but also in the move from coal power production.

*Sources:* European Council, 2010; OECD, 2016.

# 2 FINDINGS[4]

All ten European countries reviewed still provided some form of subsidy to coal in 2016.

In aggregate, these ten countries provided €6.3 billion per year in subsidies to coal (average 2005 to 2016), across a total of 65 subsidy measures (see Table 7.1). Six of the ten countries reviewed have introduced eight new coal subsidies, worth €875 million per year, since the Paris climate agreement in 2015.

Although the value of several subsidies (16 out of 65) could not be quantified (see following section), the highest level of average annual subsidies is provided by the G20 host in 2017, Germany. This includes over €2 billion in subsidies to coal mining, which Germany has committed to ending by 2018.

## 2.1 Transparency

The first step in European governments reaching their commitments is to clearly identify and estimate current subsidies, including through processes such as the G20 peer reviews (OECD, 2016).

Unfortunately, transparency of information on fossil fuel subsidies,

*Table 7.1 New and continuing subsidies to coal in Europe (average annual 2005–2016 € million)*

| Country/ subsidy category | 1. Coal mining | 2. Transition support (workers and communities) | 3. Decommissioning and environmental rehabilitation | 4. Refining and processing | 5. Capacity mechanism | 6. Biomass co-firing | 7. Coal-fired power (other) | 8. EU Emissions Trading Scheme (EU ETS) | 9. Industry | 10. Households | 11. Research and development | Country total | Count total number of current subsidies |
|---|---|---|---|---|---|---|---|---|---|---|---|---|---|
| Czech Republic | 66.2 | 8.0 | 6.4 | 0.0 | 0.0 | 0.0 | 0.0 | d.n.a. | 0.0 | 34.1 | 0.5 | **115.2** | 6 |
| France | 0.0 | 0.0 | 0.0 | 0.0 | d.n.a. | 0.0 | d.n.a. | 0.0 | 2.3 | 0.0 | 0.2 | **2.4** | 4 |
| Germany | 2,248.2 | 176.1 | 239.4 | 9.8 | 230.0 | 0.0 | 154.1 | 0.0 | 128.8 | 0.0 | 15.8 | **3,202.1** | 12 |
| Greece | 0.0 | 0.0 | 0.0 | 0.0 | 149.3 | 0.0 | d.n.a. | 0.0 | 1.3 | d.n.a. | 0.0 | **150.6** | 4 |
| Hungary | 0.0 | 28.7 | 3.4 | 0.0 | 0.0 | 0.0 | 32.7 | 0.0 | 0.0 | 8.7 | 0.0 | **73.5** | 5 |
| Italy | 0.0 | 0.0 | 0.0 | 0.0 | 0.0 | 0.0 | 0.0 | d.n.a. | 0.0 | 0.0 | 8.9 | **8.9** | 2 |
| Netherlands | 0.0 | 0.0 | 0.0 | 0.0 | 0.0 | 450.0 | 189.0 | 0.0 | 0.0 | 0.0 | 0.2 | **639.2** | 3 |
| Poland | 389.1 | 273.5 | 230.0 | 0.0 | d.n.a. | d.n.a. | 0.0 | d.n.a. | 0.0 | 9.5 | 17.5 | **919.5** | 12 |
| Spain | 283.1 | 372.8 | 15.0 | 0.0 | 83.0 | 0.0 | d.n.a. | 0.0 | 0.0 | 0.0 | 0.5 | **754.4** | 9 |
| UK | 48.6 | 0.0 | d.n.a. | 0.0 | 138.4 | 0.0 | 238.4 | 0.0 | 0.0 | 0.0 | 9.0 | **434.5** | 8 |
| TOTAL (10 countries) | **3,035.2** | **859.0** | **494.2** | **9.8** | **600.7** | **450.0** | **614.3** | **Not available** | **132.3** | **52.3** | **52.5** | **6,300.3** | **65** |

*Note:* d.n.a. = data not available.

along with accountability for phasing out those to coal remains limited, and 16 out of the 65 subsidies identified could not be quantified. Most non-quantifiable subsidies are linked to support for the use of coal for power or industry, while there is far more data available for subsidies to coal mining. This is likely linked to the EU Directive on phasing out state aid (subsidies) to uncompetitive hard coal mines by 2018.

Overall, the analysis of subsidy reporting demonstrates the significant gap between European countries in terms of their reporting on subsidies (including those to coal) (see Table 7.2). Only one country reviewed, Germany, regularly reports on its subsidies, under the biannual Subsidy Report of the Federal Government. In 2017, Germany also participated alongside Mexico in a G20 peer review process of its fossil fuel subsidies.

Italy has also completed a first-time inventory of environmentally harmful subsidies – including to fossil fuels – by launching a Catalogue of Environmental Subsidies at the end of 2016. In sharp contrast, the UK government explicitly denies that it provides any subsidies to fossil fuels, although the UK's fossil fuel subsidies have been documented by the OECD and the International Monetary Fund.

## 2.2 Coal Mining – Subsidy Phase-out

Subsidies to coal mining still represent the highest proportion of total coal subsidies provided across the ten countries (see Table 7.1, category 1 – 48 per cent by value). This may partly be attributed to increased transparency in reporting. These subsidies are increasingly framed to represent a shift away from supporting ongoing coal mining and facilitating mine closures. Linked to the EU Directive on phasing out state aid to uncompetitive hard coal mines by 2018, this is also coupled with the fact that many coal mines in Europe are no longer making profits. In addition, 8 per cent of total subsidies to coal by value (€494 million per year) across the ten countries are directed towards the decommissioning and environmental rehabilitation of coal mines (see Table 7.1, category 3).

Countries are found to be at different stages in the phase out of subsidies to coal mining (see Table 7.3). Many remaining subsidies to coal mining are focused on closing coal mines (e.g. in the Czech Republic, Germany, Greece, Poland and Spain). Though its mining industries are far smaller, the UK still provides various tax benefits to ongoing coal mining (see Box 7.2).

## 2.3 Coal-fired Power – Subsidy Phase-out

In contrast to the shift in support to coal mining heading towards closure, many of the new subsidies for coal-fired power being introduced risk

*Table 7.2 Scoring of countries by transparency of coal subsidy reporting*

| Country | Czech Republic | France | Germany | Greece | Hungary | Italy | Netherlands | Poland | Spain | UK |
|---|---|---|---|---|---|---|---|---|---|---|
| Transparency | Poor | Poor | Very good | Poor | Poor | Good | Poor | Poor | Good | Very poor |

*Notes:* Scoring criteria
Very good: regular reporting of fossil fuel subsidies (i.e. annually or biannually), including those to coal and participation in fossil fuel subsidy peer review.
Good: reporting of fossil fuel subsidies, including those to coal, although not on a regular basis (i.e. annually or biannually).
Poor: no reporting of fossil fuel subsidies, including those to coal, except for in international inventories, including those of the OECD and IMF.
Very poor: statement denying provision of fossil fuel subsidies, including those to coal, despite documentation in international inventories, together with those of the OECD and IMF.

*Table 7.3 Scoring of countries by progress in phasing out subsidies to coal mining (see categories 1, 2 and 3 in Table 7.1)*

| Country | Czech Republic | France | Germany | Greece | Hungary | Italy | Netherlands | Poland | Spain | UK |
|---|---|---|---|---|---|---|---|---|---|---|
| Coal mining – subsidy phase-out | Good | Not applicable – no coal mining | Good | Good | Good | Good | Not applicable – no coal mining | Good | Good | Poor |

*Notes:* Scoring criteria
Good: subsidies are only provided with the stated objective of supporting the phase out of coal mining activities, including the transition of workers and communities.
Poor: subsidies are provided to ongoing coal mining activities.

## BOX 7.3 CAPACITY MECHANISMS

With renewables accounting for an increasing share of electricity generation, many governments have become concerned about the ability to balance supply and demand when the sun is not shining and the wind is not blowing. In response, a renewed interest in 'capacity mechanisms', which offer extra payments to operators that can either turn up their supply or turn down upon demand, has emerged.

Although they may appear to provide a solution for governments seeking to ensure security of supply, they have also tended to result in large payments to fossil fuel-fired generation (including to coal plants that would otherwise be uneconomic).

For example, the UK's annual capacity market auction has received criticism for overestimating future supply needs, favouring fossil fuels and delaying coal-plant decommissioning. Germany has plans to establish a capacity reserve under which 2.7 GW of coal-fired generation will receive (currently undefined) payments for staying available as backup capacity until 2021. A planned new capacity reserve is currently under in-depth investigation by the European Commission. Meanwhile, Poland plans to spend over €20 billion to finance the creation of a capacity market, with members of the government openly discussing which coal-fired power plants would be financed.

*Sources:* van der Burg and Whitley, 2016; Littlecott, 2014; ENTSOE, 2015; ClientEarth, 2016; European Commission, 2017; Zasuń, 2014.

lengthening the life of assets. This comes despite several countries' commitments to phase out coal-fired power (see Box 7.1).

Some of the measures with a stated objective to support the energy transition to lower-carbon sources and efficiency are in fact facilitating the use of coal (€1 billion per year). This includes subsidies provided via capacity mechanisms (see Box 7.3), support to biomass power, and the EU ETS (see Table 7.1 – categories 5, 6 and 8).

These new subsidies can undermine measures (e.g. carbon price support in the UK) that are meant to increase the cost of coal-fired power to achieve emission reduction objectives. In some cases, such as in Germany, initial plans for measures that were meant to ensure that the polluter pays have eventually succumbed to pressure from companies and been replaced by subsidies that do the opposite and pay the polluter.

The European Commission noted that careful design of capacity mechanisms is needed to ensure they do not 'contradict the objective of phasing out environmentally harmful subsidies, including for fossil fuels' (European Commission, 2014). To this end, the Commission has suggested banning new power plants that emit more than 550 g $CO_2$/kWh from participating in these schemes (Rocha et al., 2017). Such a limit should be introduced by all European governments for both existing and planned capacity mechanisms and coal plants, with immediate effect, as it would effectively prohibit coal-fired power plants from profiting from these schemes.

Both Article 10c and the planned Modernisation Fund under the EU's Emissions Trading System (ETS) are aimed to be transitional mechanisms for supporting the transformation of energy systems. However, in the current phase of the ETS, a significant proportion of Article 10c support has been used to finance the retrofits of coal capacity in Central and Eastern Europe. In terms of ending coal subsidies through the EU ETS, a strict emissions performance standard should also be applied to the Modernisation Fund and Article 10c. This will ensure that these instruments will be used to support investments to energy efficiency and renewable energy sources rather than coal.

In contrast to variations in support to coal mining, as well as progress in terms of subsidy transparency, European countries seem to be on a fairly level playing field when it comes to progress in phasing out subsidies to coal-fired power, along with other uses of coal by industry and households (see Table 7.4). Overall, all ten countries reviewed continue to subsidise coal use in some form, with progress in phasing out subsidies to coal-fired power being hampered by new subsidies to coal, introduced under the auspices of the energy transition.

### 2.4 Supporting a Just Transition Away from Coal

Under the Paris Agreement, all countries have committed to 'taking into account the imperatives of a just transition[5] of the workforce and the creation of decent work and quality jobs in accordance with nationally defined development priorities'. In contrast to these pledges, we have found that only a small minority of coal subsidies by value (14 per cent) are supporting workers and communities to this end.

Subsidies with the stated objective of supporting the transition of workers and communities were found to be higher in scale in countries where coal has historically been significant to the economy (e.g. Germany, Hungary and Poland).[6] However, there is generally limited information on spending, and it is difficult to distinguish in some cases between the proportion of funds dedicated to decommissioning and rehabilitation of mine sites, and that dedicated to communities and workers.

## 3 CONCLUSIONS AND RECOMMENDATIONS

To achieve Paris Agreement climate targets, fight air pollution and protect health, as well as support a just transition to low-carbon energy

*Table 7.4 Scoring of countries by progress in phasing out subsidies to coal-fired power (see categories 5, 6, 7 and 8 in Table 7.1)*

| Country | Czech Republic | France | Germany | Greece | Hungary | Italy | Netherlands | Poland | Spain | UK |
|---|---|---|---|---|---|---|---|---|---|---|
| Coal-fired power – subsidy phase-out | Poor | Very poor | Very poor | Very poor | Very poor | Poor | Poor | Very poor | Poor | Poor |

*Notes:* Scoring criteria
Good: no subsidies to coal-fired power identified.
Poor: subsidies are provided in the context of the transition away from coal-fired power, namely for plant or facility upgrades, or to compensate plants as part of a closure plan.
Very poor: subsidies are provided to ongoing coal-fired power production.

systems, European countries will need to rapidly phase out coal. Driven by sharp reductions in the cost of renewable energy technologies, effective campaigns and legal action by civil society groups, governments are implementing various measures to make this happen. However, at the same time, they are offering new subsidies that provide a lifeline to coal.

There are three key areas that European governments must focus on to achieve a complete phasing out of coal subsidies:

- Governments in Europe must ensure that mechanisms with the stated focus of supporting the energy transition do not support coal. This includes ending subsidies for coal under the EU's ETS, capacity mechanisms, and subsidies to biomass power generation.
- Any remaining subsidies must be focused on supporting a just transition for workers and communities – ensuring that companies and utilities also meet their obligations.
- The above must be supported by increased transparency and accountability to meet existing subsidy phase-out commitments – with all governments undertaking consistent annual reporting of subsidies to coal and other fossil fuels.

It is clear that countries in Europe are at the forefront of moving towards zero-carbon energy systems (see Box 7.4). To achieve this goal, all government resources must be used to accelerate the energy transition, not slow

### BOX 7.4 IS THE TRANSITION AWAY FROM COAL ACHIEVABLE?

Recent analysis by Carbon Tracker and the Grantham Institute has found that solar photovoltaics (PV) (with associated energy storage costs included) could supply 23 per cent of global power generation in 2040 and 29 per cent by 2050, allowing for a complete phasing out of coal and leaving natural gas with just a 1 per cent market share (Sussams and Leaton, 2017).

In Europe, both Germany and the UK have shown that electric grids can cope well with a coal phase-out. In the UK, coal use has declined substantially in recent years without any disruption to security of supply; as the remaining coal plants are taken offline, the challenge for government is to encourage more investment in renewables, energy efficiency, storage and demand-side management (Wynn and Schlissel, 2017).

New data from Wind Europe indicates that renewable energy sources made up nearly nine-tenths of new power added to Europe's electricity grids last year, with wind power outstripping coal to become the EU's second-largest form of power capacity (Vaughan, 2017).

it down. European governments must demonstrate global leadership and this must begin with phasing out subsidies to coal, followed by an end to all fossil fuel subsidies.

## ACKNOWLEDGEMENTS

This material was funded by the Oak Foundation and the Hewlett Foundation. The authors are grateful for support and advice on this policy briefing from Chris Littlecott (E3G), Dave Jones (Sandbag), Joanna Flisowska (Climate Action Network Europe), and Andrew Scott (ODI). The authors would also like to thank Holly Combe and Amie Retallick for editorial support.

## NOTES

1. Both RWE and E.ON have divided themselves in two, creating respectively the entities Innogy and Uniper. RWE and Uniper have the old gas- and coal-fired power stations, while E.ON and Innogy hold the clean, green businesses such as infrastructure and renewables (Chazan, 2016).
2. The European Semester provides a framework for the coordination of economic policies across the EU, allowing countries to discuss their economic and budget plans and monitor progress at specific times throughout the year.
3. Although not included in our analysis, subsidies are also provided to coal in Europe through a broader range of instruments than budgetary support and tax breaks. This includes investment by state-owned enterprise and public finance, such as two new lignite power plants in Greece receiving support through Public Power Corporation (PPC), with one of the plants being underwritten by a loan from a consortium led by KfW-Ipex, the German public export credit agency.
4. These estimates do not include historic subsidies that have been phased out, which are discussed within each country brief, but not included in totals.
5. Per the International Trade Union Confederation, a just transition brings together workers, communities, employers and government in social dialogue to drive the concrete plans, policies and investments needed for a fast and fair transformation. This focuses on jobs, livelihoods and ensuring that no one is left behind as we race to reduce emissions, protect the climate and advance social and economic justice.
6. For context, the European Association for Coal and Lignite (EURACOAL) estimates that in 2015, coal mining employed 185,000 people across the region, including some at integrated mine and power plants. This is 0.08 per cent of the EU's total workforce, which is currently estimated at 220 million (EURACOAL, 2016; Eurostat, 2017).

## REFERENCES

Alberici, S., Boeve, S., van Breevoort, P., Deng, Y., Förster, S., Gardiner, A., van Gastel, V., Grave, K., Groenenberg, H., de Jager, D., Klaassen, E., Pouwels, W.,

Smith, M., de Visser, E., Winkel, T. and Wouters, K. (2014) 'Subsidies and costs of EU energy: final report'. Brussels: European Commission.
Chazan, G. (2016) 'Eon and RWE pursue radical restructurings'. *Financial Times.*
Clark, P. (2017) 'Dong aims to phase out coal-fired power generation by 2023'. London: *Financial Times.*
ClientEarth. (2016) 'Poland's support for coal through capacity market will hit household bills.' Warsaw: *Client Earth.*
Crisp, J. (2017) '23 EU countries are breaking European air quality laws'. London: *EurActiv.*
DeSmogBlog. (2016) 'Europe Quits Coal'. Cleveland: *Ecowatch.*
EURACOAL. (2016) 'Coal industry across Europe'. Brussels: *EURACOAL.*
European Commission. (2011) 'A roadmap for moving to a competitive low carbon economy in 2050', COM (2011) 112. Brussels: European Commission.
European Commission. (2014) 'Guidelines on State aid for environmental protection and energy 2014–2020', (2014/C 200/01). Brussels: European Commission.
European Commission. (2016) 'Energy prices and costs in Europe'. Brussels: European Commission.
European Commission. (2017) 'State aid: Commission opens in-depth investigation into German plans for electricity capacity reserve'. Brussels: European Commission.
European Council. (2010) Council Decision of 10 December 2010 on State aid to facilitate the closure of uncompetitive coal mines.
European Network of Transmission System Operators for Electricity (ENTSOE). (2015) *ENTSO-E Annual Report 2015*. Brussels: ENTSOE.
Eurostat. (2017) *Employment and unemployment (LFS).*
G7. (2017) 'Chair's Summary: G7 Rome Energy Ministerial Meeting'. 9–10 April.
G20. (2016) 'G20 Leaders' Communique Hangzhou Summit'. 4–5 September.
International Energy Agency (IEA). (2016) *IEA 2016 World Energy Outlook*. Paris: IEA.
IPCC (Intergovernmental Panel on Climate Change). (2014) 'Climate Change 2014 Synthesis Report'. Geneva: Intergovernmental Panel on Climate Change.
Jones, D., Huscher, J., Myllyvirta, L., Gierens, R., Flisowska, J., Gutmann, K., Urbaniak, D. and Azau, S. (2016) *Europe's Dark Cloud – how coal burning countries are making their neighbours sick*. Brussels: CAN Europe; Brussels: European Environmental Bureau (EEB); Brussels: HEAL; Brussels/London: Sandbag; Gland: WWF.
Littlecott, C. (2014) 'Keeping coal alive and kicking: Hidden subsidies and preferential treatment in the UK Capacity Market'. London: E3G.
Littlecott, C. (2017) 'Summary: The coal phase out transition – Italy's leadership opportunity'. Washington D.C.: E3G.
Madson, D. (2017) 'Finland is abandoning coal'. New Haven: Yale Climate Connections.
McGlade, C. and Ekins, P. (2015) 'The geographical distribution of fossil fuels unused when limiting global warming'. *Nature* 517: 187–90.
OECD. (2016) *OECD-IEA Fossil Fuel Support and Other Analytics: Publication*. Paris: OECD.
Rocha, M., Parra, P.Y., Sferra, F., Schaeffer, M., Roming, N., Ancygier, A., Ural, U. and Hare, B. (2017) 'A stress test for coal in Europe under the Paris Agreement'. New York: Climate Analytics.
Sartor, O. and Spencer, T. (2015) 'Fossil fuel subsidies and the new EU Climate and

Energy Governance Mechanism'. Paris: Institute for Sustainable Development and International Relations (IDDRI).
Shearer, C., Ghio, N., Myllyvirta, L., Yu, A. and Nace, T. (2017) *Boom and Bust 2017: Tracking the Coal Plant Pipeline*. CoalSwarm, Greenpeace USA, and Sierra Club.
Sussams, L. and Leaton, J. (2017) *Expect the unexpected: The disruptive power of low-carbon technology*. London: Carbon Tracker and Grantham, Imperial College.
United Nations (UN). (2015) 'Sustainable development knowledge platform'. New York: UN.
UNFCCC. (2015) 'Paris Agreement to the United Nations on Climate Change', UN Doc FCCC/CP/2015/L.9. 12 December.
van der Burg, L. and Whitley, S. (2016) *Rethinking power markets: capacity mechanisms and decarbonisation*. London: ODI.
Vaughan, A. (2017) 'Almost 90% of new power in Europe from renewable sources in 2016'. London: *The Guardian*.
World Resources Institute. (2015) 'CAIT Climate Data Explorer'. Washington D.C.: World Resources Institute.
Wynn, G. and Schlissel, D. (2017) *Electricity-Grid Transition in the U.K.* Cleveland: Institute for Energy Economics and Financial Analysis (IEEFA).
Zasuń, R. (2014) 'Euroobligacje uratują Kompanię Węglową?' Warsaw: WysokieNapiecie.pl.

# 8. Noise pollution taxes: a possibility to explore

**Marta Villar Ezcurra**

## 1 INTRODUCTION

Noise pollution has wide-ranging adverse health, biodiversity, social and economic effects. It is more severe and widespread than ever before, and will continue to increase worldwide because of mechanisation, urbanisation and population growth.[1] Noise from road traffic alone is the second most harmful environmental stressor in Europe, behind air pollution, according to the World Health Organization (WHO).[2]

Legal measures applied at different levels of government from local authority bylaws through national regulations to EU Directives include effect-oriented and source-oriented instruments and involve, among other concerns, the control of noise emissions through emission standards for road and off-road vehicles; emission standards for construction equipment and emission standards for plants.[3] Noise control requirements in European countries are typically determined from the effects of noise on health and the environment (effect oriented). Other countries base their noise management policies on the requirement for best available technology or techniques that do not entail excessive cost (source oriented) (e.g. for aircraft noise).

One of the possible ways of addressing noise pollution is through the use of environmental taxes as a complement to $CO_2$ taxes. In this regard, the Mirrlees Review[4] proposes to focus on two priorities: greenhouse gas (GHG) emissions and road congestion. Successful noise management should be based on the fundamental environmental principles of precaution, polluter pays (PPP) and prevention.[5] A good design of environmental taxes addressing the problem of noise pollution is also required to ensure their effectiveness.[6] To that end, it is essential to measure the environmental damage caused by noise;[7] to define the scope and sectors (e.g. for industrial, transportation or airport); to calculate the associated health costs or damage to biological diversity, and the relationship of acoustic emissions with $CO_2$ emissions in the case of engines; as well as seeking solutions in the sources of noise (promoting silent or insulating devices) and in the means of transmission (engines) or of receivers.

Currently in Europe, the maps resulting from Directive 2002/49/EC (END)[8] make it possible to better assess the impact on public noise policy and design priorities in noise management. However, only some EU Member States use taxes on air transport for noise pollution as such to mitigate sound levels or to finance compensatory measures (as in the case of the French 'Tax on Air Transport Noise Pollution' (TNSA), payable by all aircraft operators irrespective of nationality) and most of the countries use taxes in relation to road traffic. China has also recently adopted an important environmental tax regulation, while in the United States, tax incentives have been introduced by some States, like Kentucky and Ohio, specifically labelled as such to address noise pollution (see Sections 4.3 and 5).

This chapter explores the current state of noise pollution taxation in order to identify what role it can and should play within the environmental taxation framework.

## 2 THE ENVIRONMENTAL RELEVANCE OF NOISE AS A PREMISE FOR PROPOSING NEW TAXES

Some scholars remark on quite distinct variations in distinguishing how 'noise' and 'environmental noise' are defined, pointing out that environmental noise, as a form of pollution, is 'something that is to be avoided, controlled, regulated or eliminated because of its negative impact on humans and human-environment relations'.[9] Therefore, among different market-based instruments (MBIs) there is room for considering environmental taxes in combating noise pollution. Even so, while there is a growing public awareness in the fight against air and water pollution, this third jeopardy (noise pollution) receives less attention from policymakers and public opinion than the other two types. Noise is the forgotten environmental problem.

The adverse impact of noise pollution has been well documented for a long time. According to WHO data from 2012, one in five Europeans is regularly exposed to sound levels at night that could significantly damage health. Indeed, new evidence has emerged showing that at least 1 million healthy life years are lost every year in Europe as a result of noise from road traffic alone.[10] Moreover, the social costs of traffic noise across 22 EU states are over €40 billion per year, with passenger cars and lorries responsible for the bulk of these costs.[11]

Exposure to noise is a major health concern in urban areas.[12] A recent study[13] found that Guangzhou in China had the worst noise pollution, while Zurich in Switzerland had the least. Barcelona, one of only two European cities to feature in the worst ten, came seventh, while capital cities Mexico City, Paris and Buenos Aires came in at eighth, ninth and tenth position respectively.[14] Examples of local measures include installation of road and rail noise barriers, managing flight movements around airport locations and reducing noise at source. However, according to the European Environmental Agency (EEA), further efforts will require an updated noise policy aligned with the latest scientific knowledge, as well as improvements in city design, being the most effective actions that reduce noise at its source, for example by decreasing noise emissions of individual vehicles by introducing quieter tyres. Green areas can also assist in reducing urban noise levels (e.g. rethinking urban design).[15]

Nevertheless, it would be wrong to assert that noise is an environmental problem that is exclusive to cities. Environmental noise is a global issue and does not recognise national borders. Thus, strategies are needed to address noise mitigation beyond core city regions. Economic measures and MBIs can help. In particular, environmental taxes can address noise pollution following environmental principles such as the PPP. According to the EEA 2016 Report, using general and mixed MBIs, EU Member States must adopt transport and noise action plans to manage noise issues and effects for all major agglomerations, airports, roads and railways. These plans may include regulatory or economic measures and incentives.[16]

### 2.1 Correlation Between Environmental Noise and Air Pollution Exposure

There is evidence that environmental noise is related to air pollution, leading to a greater impact on human health.[17] This factor is essential for optimal design of environmental noise taxes. The range of correlations between environmental noise and air pollution indices shows the highest correlation is in relation to road traffic sources.[18] This highlights the value of considering integrating mitigation approaches that address common sources of both air pollution and noise, such as road transport (see Section 3.1). For road traffic, noise is mainly produced by the engine and by the contact of tyres on the ground, while air pollution consists of engine exhaust.[19] In the case of aircraft, noise may arise from the engines but also from the aircraft structure and is most prominent for landing and take off, as well as the noise of the aircraft on the ground. On the other hand, air pollution from aircraft is the result of emissions from the plane's engines. Rail noise arises from the contact of the train wheels with the track, from the locomotive engine, and from wind resistance to the train and is often accompanied by vibration. Air pollution is caused by the emissions from the train's engine.

### 2.2 Identifying the State and Sources of Noise in Europe

Although railways,[20] airports[21] and industry[22] are important sources of environmental noise, road traffic noise, both inside and outside urban areas, is still the predominant source affecting human exposure: at least 100 million people are exposed to levels of traffic noise that exceed the European Union's indicator of noise annoyance and measures indicate levels above the 55 dB Lden action levels defined by the END. Where data has not been reported by Member States, gap-filling has been performed.[23]

### 2.3 The Factor of the Relationship Between Noise Pollution and Fundamental Rights

Noise emission and noise immission are terms often incorrectly interchanged. Noise emission is emitted by the source while noise immission is the noise experienced by individuals. Thus, noise emission is dependent on properties of the source, while noise immission is dependent on everything between source and receiver (that is multiple sources – the presence of obstacles, the ground cover and so on).[24] Usually, there should be enough legal protection for individuals, who can require public authorities to react by punishing the immission noise.

In some cases, courts have identified (on a case-by-case basis) instances which warrant the protection of fundamental rights against the immisions noise. The European Court of Human Rights has on several occasions examined the relationship between serious environmental pollution and the right to respect for private and family life and the home.[25] The Court has given clear confirmation that Art. 8 of the European Convention on Human Rights guarantees the right to a healthy environment. It found violations of Art. 8, on both occasions unanimously, in *López Ostra v. Spain* and in *Guerra and Others v. Italy*.[26] In *Fadeyeva v. Russia*[27] the Court stressed the importance of the effective measures to be applied, which consider the interests of the local population affected by the pollution, and 'which would be capable of reducing the industrial pollution to acceptable levels'. Indeed, in *Hatton and others v. the United Kingdom*[28] the Court pointed out that one of the important functions of human rights protection is to protect 'small minorities' whose 'subjective element' makes them different from the majority.[29] This judgment did not consider the international standards concerning the effects noise has on sleep, although the relevant data was available in the file.[30]

This positive approach to fundamental rights has been followed by national courts. Among others, the Spanish Constitutional Court recognises that the mandate to protect human health (Art. 43 of the Spanish Constitution) and the environment (Art. 45 of the Spanish Constitution) encompass in scope the protection against noise pollution. In addition, some fundamental rights recognised by the Constitution, among others, the right to personal and family privacy, enshrined in Art. 18(1), also served as constitutional support.[31]

## 3 STRATEGIC NOISE MAPPING: THE RELEVANCE OF EU DIRECTIVE 2002/49/EC (END)

Although similar approaches have been applied outside the EU, they have not been completed within the context of a strategic plan for noise reduction but as ad hoc measures to reduce noise. EU legislation has moved from being almost entirely focused on regulating noise at its source, to attempting to mitigate environmental noise at the point of the receiver through the establishment of the 2002 END.[32]

The purpose of the END is to define a common approach intended to avoid, prevent or reduce on a prioritised basis the harmful effects, including annoyance, of exposure to environmental noise.[33] To that end, EU Member States were required to assess exposure to noise from key transport and industrial sources with two initial reporting phases: 2007 and 2012. Member States must produce strategic noise maps for all major roads, railways, airports and agglomerations on a five-year basis. Where the recommended thresholds for day and night indicators are exceeded, action plans must be implemented.

In Europe, a common approach for assessing noise levels is necessary in order to improve CNOSSOS-EU (Common Noise Assessment Methods),[34] which is used for noise mapping. Currently, there are two key END indicators: Lden (day evening and night exposure) and Lnight (night time exposure) and 2012 strategic noise maps reported are presented, as well as calculations for annoyance and sleep disturbance, hospital admissions and mortality.[35] What appears certain is that the END application experience has led to improvements in the process, both in rclation to the mapping criteria and to the procedures for the delivery of results. In 2002 a noise map for the agglomeration of Madrid was made based on 4,395 measuring points. This led to the development of a new measurement system, known as the SADMAM,[36] which complied with the Directive in a more effective manner.

### 3.1 Road Traffic Noise Source Emission as the Main Source of Noise Pollution

Road traffic as a source of noise pollution is determined by combining the noise emission of each individual vehicle within a traffic flow. To that aim, vehicles are separated into four distinct categories which take into account varying levels of noise emission: (i) light motor vehicles; (ii) medium heavy vehicles; (iii) heavy vehicles; (iv) powered two-wheelers.[37] Moreover, the road gradient has two effects on the noise emission of the vehicle: first, it affects the vehicle speed and thus the rolling and propulsion noise emission of the vehicle; second, it affects both the engine load and the engine speed via the choice of gear and thus the propulsion noise emission of the vehicle. The effect of the acceleration and deceleration of vehicles and the type of road surface are also key factors. Environmental benefits from hybrid electric and pure electric vehicles have resulted in a substantial reduction of the noise emitted by such vehicles.

### 3.2 Air Traffic, Railway Traffic and Industrial Noise Sources Emission

Noise pollution from aircraft is subject to special treatment due to the significant methodological requirements imposed by Directive 2002/30/EC. Conversely, various sources like depots, loading/unloading areas, stations, bells or station loudspeakers are associated with railway noise. These sources are to be treated as industrial noise (fixed noise sources). Switzerland has one of the world's most advanced programmes of noise abatement for railways. The programme was funded through taxes on heavy vehicles, VAT and fuel taxes, and through the capital markets.[38]

## 4 IS IT THE RIGHT TIME TO PROMOTE ENVIRONMENTAL TAXES TO FIGHT AGAINST NOISE POLLUTION?

Currently, one of the biggest challenges for many governments – at least in the EU – is the reduction of major fiscal deficits with the least collateral damage to the economy. Environmental taxation is high on the political agenda. Although the primary objective of environmental taxes is not to raise public revenues, but rather to tackle environmental challenges, some countries have implemented 'environmental tax reforms'.[39] The effect of these policy instruments is well documented in both economic and political literature.[40] The current application of environmental taxes in EEA countries shows that energy taxes represented by far the highest share of overall environmental tax revenue accounting for 76.7 per cent of the EU-28 total in 2015 while taxes on pollution and resources reached only 3.5 per cent of the EU-28 total.[41]

As the European Commission has pointed out in its Communication of 26 March 1997,[42] environmental taxes and charges can be an appropriate way of implementing the PPP, by including environmental costs in the price of goods or services.[43] An overview of environment-related taxes in EU and EEA countries as of October 1996 indicates that there are very few examples of noise charges (in Belgium, France, Germany, the Netherlands, Norway, Portugal and Sweden) and they were only implemented within the air transport sector.[44] Although the number of environmental taxes has indeed increased over the past decade and the redesign of existing ones may have helped improve their effectiveness, noise pollution charges and taxes – at least labelled as such – still remain generally almost unchanged.

In any case, taxation on noise pollution imposes no inconvenience from the theoretical point of view, but as experience shows, it continues to be symbolic in practice. On the other hand, it was noted that, while efficiency considerations do not require pollution taxes to be used to compensate victims, actual legislation, existing and proposed, tends to operate with a 'revenue requirement' for this purpose.[45]

### 4.1 Taxes on Road Traffic Noise Pollution

For traffic noise, some scholars propose a 'traffic noise rating'.[46] The zone rating is introduced on the assumption that, as a rule, the noise impact of vehicles owned by urban residents will be greater than that of vehicles owned by rural or semi-rural residents. This assumption is questionable for heavy commercial vehicles and for the heavier types of private cars since the application of a mileage coefficient is necessary.[47] Other scholars have raised the question as to whether there will be a 'noise tax', paid at the time of purchase, on cars and motorcycles that make excessive noise.[48]

For transport and noise, the Eurovignette Directive,[49] which is fiscal legislation, establishes common rules on distance-related tolls and time-based user charges (vignettes) for the use of certain infrastructure by heavy goods vehicles. According to recent amendments to the Directive, Member States may maintain or introduce external-cost charges related to the 'cost of traffic-based air pollution' – the cost of the damage caused by the release of particulate matter and ozone precursors during operation. The revenues generated from external-cost charges should be used to make transport more sustainable.[50] Apart from that, transport tax design varies widely between countries. It includes both one-off taxes such as sales/registration taxes and recurrent ones such as annual circulation taxes. Sales taxes have been implemented in fewer European countries than circulation taxes: 21 of 28 EU Member States compared to 28 out of 28. Private vehicles, however, are not subject to annual circulation taxes in all countries: 6 out of 28 EU Member States do not levy them. In recent years, road user charges have become more common for private as well as commercial vehicles.[51]

### 4.2 Taxes on Air Transport Noise Pollution

Emissions, noise pollution and congestion all provide economic rationales for aviation taxes. International agreements prevent fuel for international flights being taxed. But taxes on tickets, passengers and flights are all permissible. For aircraft noise charges and taxes can be calculated more accurately than in other sources of noise cases, since the noise levels and noise 'footprints' of each aircraft are well known.[52] Various studies have attempted to quantify the marginal external costs of aviation and thus the appropriate size of an aviation tax. Some have focused exclusively on the emissions externality while others examined both emissions and noise externalities.[53] Depending on the technology and scenario used, the average 'external' cost of air travel is about EUR 0.01 to EUR 0.05 per passenger-kilometre.[54] The move from a per-passenger to a per-flight tax in 2009 now has cross-party support. Such a move may provide considerable environmental benefits if it can be designed to target reasonably accurately the various externalities involved, without causing too much administrative complexity.

Nowadays, there are several aviation taxes in different countries,[55] but very few directly address noise pollution.[56] In transport taxes (e.g. in Austria and France), each aircraft owner is required to pay an air transport levy to the competent tax office for passengers departing from national airports; airline/air carrier taxes (e.g. Croatia and Italy), levy air carriers or airlines; passenger taxes (e.g. Germany, Portugal, UK,[57] Italy and the Netherlands) are levied on passengers flying outside the country or tax on air carriers per passenger; development and solidarity taxes (e.g. France) levy an additional tax to be allocated to the Solidarity Fund for Development.

Looking into the future, these sources of noise pollution are a natural target for international taxation. Another advantage is the likely buoyancy of a stable tax base. With projections for high and stable future growth in international airline traffic, generally exceeding that for economic activity as a whole, this will mitigate any possible dampening effect that these taxes may have.

Taxes on air transport noise pollution in France[58] are a good example of an environment-related tax.[59] Unlike the Spanish landing fees,[60] France uses these taxes mainly to provide financial assistance to the neighbourhood impacted by the proximity of the aerodromes.[61] However, the law regulating the tax provides for other use of revenues:

> in case of need, in a limit of half of the annual tax proceed, to the reimbursement of loan annuities to public entities for the financing of works aimed at reducing noise pollution; and to the reimbursement of the sums advanced granted for the financing of the works aimed at reducing noise pollution .[62]

One of the key considerations of any noise mapping assessment is identifying where transportation noise ends and industrial noise begins. In that regard, noise from ground operations is not generally considered in the noise prediction of aircraft noise. Noise from ground operations can range from the taxiing of the aircraft to the runway, servicing and related activities associated with the running of an airport (such as baggage handling) and other sources of industrial noise.

### 4.3 Taxes on Industrial Noise

A particular example in this category is China, where industrial noise is taxed in five categories (from 1–3 dB above standard to 16 dB or more above standard). In the first category, industrial noise is levied where there are multiple spots with noise levels above standard along the border of one institution, being the tax rate calculated based on the noise level at the highest spot; where there are two or more spots with noise level above standard along a border of 100 meters or more the tax rate payable is calculated as for two institutions.[63] In Singapore, to facilitate effective measures to counter noise hazards, an incentive to enjoy 100 per cent depreciation in the first year of purchase was introduced. However, it was only granted if certain qualifying criteria relating to noise levels were satisfied. This incentive is applied to expenditure incurred from 1 January 1998 onwards.[64] Other countries like Serbia have an accelerated depreciation of fixed assets serving for noise control.[65] On the other hand, noise pollution from wind turbines is not considered when developing strategic noise maps under END. While wind turbine or air conditioner noise are usually much lower than noise from other industrial sources, detailed assessment of the background noise is required prior to the development of a wind farm or air conditioner equipment.

## 5 TAX INCENTIVES RELATED TO NOISE POLLUTION AS A PATH FOR A MITIGATION APPROACH

Without any doubt, the most effective noise control and regulation measures are those that target a reduction in noise emitted at the source. By far the most effective and cost-efficient method of reducing noise is via regulation which sets out permissible noise levels. However, tax incentives related to the protection of the environment can also be an effective tool.

Most developed countries include in their tax regulations different types of environmental incentives, including those related to the elimination or reduction of noise pollution. They vary from property tax benefits, to sales tax exemptions or reductions, to corporate income tax incentives usually related with fixed assets, equipment or facilities addressed to mitigating noise pollution.

Some countries have specific incentives for noise-reduction purposes independent of other environmentally driven tax benefits. In the USA, Kentucky and Ohio are good examples of particular interest. In Kentucky, all local governments must develop, adopt and maintain a comprehensive programme of noise regulation.[66] Tax incentives for pollution control equipment include exemption from the 6 per cent state sales and use tax for facilities and equipment installed to eliminate noise pollution and certified as pollution control facilities. Certified pollution control facilities are exempt from all local property taxes.[67] Similar tax incentives are available in Ohio.[68] At the local level, many cities have clear objectives in their plans to act on noise pollution but do not include any specific tax measures. For instance, it is common to promote the use of machinery that has updated sound power certification, to ensure low noise emission, but there are no tax benefits to promote changes in behaviour related to noise pollution. Therefore, there is clearly much more that can be done through the use of tax incentive instruments.

## 6 CONCLUSIONS

There is a common understanding among international organisations, health authorities and scholars about the extremely negative effects of noise pollution and the costs associated with it. However, the diversity of noise sources and the difficulties in measuring noise pollution make public management complex. At present, legal actions setting standards for the reduction of noise at source are insufficient. Thus, although politically complicated, it is time to promote a new approach and noise pollution taxes are a feasible option.

This chapter has shown that although there are few environmental taxes in force directly addressing noise pollution, in instances where these have been applied, they have been successful. Therefore, it is arguable that their fundamental structure could easily be replicated in other jurisdictions.

Nowadays, certain circumstances favour noise pollution taxes becoming a real option for environmental noise protection: the technology of motor vehicles is in a state of development, with the emergence of the electric car a real possibility for mass use, allowing for new taxes aiming to drive the market towards options that are more favourable to the environment with respect to noise emissions and which, simultaneously, help neutralise the social costs of acoustic emissions.

Although noise pollution taxes can be applied in several cases and have a tremendous flexibility, which allows them to be applied at the state, regional or local level, depending on the characteristics of each country, their application to motor vehicles is of particular interest, due to the number of motor vehicles and their polluting effects, and because of the growing concentration of the population in large cities.

A further advantage of noise pollution taxes is their easy implementation alongside taxes designed to combat the emission of GHGs. Emitters of noise and pollutant gases, especially in the case of motor vehicles, are co-existent and measures to reduce noise emissions are, in most cases, also effective in reducing emissions of $CO_2$ and other harmful gases.

## NOTES

1. See Goines, L. and Hagler, L. (2007), Noise pollution: a modern plague, *Southern Medical Journal* 100(3), 287–94.
2. See Stansfeld, S.A. (2015), Noise effects on health in the context of air pollution exposure, *Int J Environ Res Public Health*, 14 Oct 12(10), 12735–60, doi: 10.3390/ijerph121012735; WHO (2011), *Burden of disease from environmental noise. Quantification of healthy life years lost in Europe*, Copenhagen, available at: http://www.euro.who.int/__data/assets/pdf_file/0008/136466/e94888.pdf, accessed 23 January 2018.
3. Noise pollution is primarily a local problem, but one which calls for a Union-wide solution.
4. See Mirrlees, J., Adam, S., Besley, T., Blundell, R., Bond, S., Chote, R., Gammie, M., Johnson, P., Myles, G. and Poterba, J. (2011), The Mirrlees Review: Conclusions and Recommendations for Reform, *Fiscal Studies* 32(3), 331–59, 0143-5679, p.339, doi:10.1111/j.1475-5890.2011.00140.x.
5. The United Nations' Agenda 21 (UNCED 1992) and the European Charter on Transport, Environment and Health (London Charter 1999) both support many environmental management principles on which government policies, including noise

management policies, can be based. See, http://www.noiseoff.org/document/comnoise.5.pdf, accessed 23 January 2018.

6. Although there are different definitions of the concept 'environmental tax' (e.g. OECD, EEA) which focus on the nature of the tax base, there is no inevitable connection between taxing a polluting tax base and obtaining environmental effects. Choosing the tax base is due to it being the only objective way of identifying and comparing tax data internationally. See Eurostat (2013), *Environmental taxes. A statistical guide*, Luxembourg: Publications Office of the European Union, doi:10.2785/47492.
7. Whatever method applies should follow an international or national standard. Most national agencies have guidelines that should be adhered to.
8. See Directive 2002/49/EC of the European Parliament and of the Council of 25 June 2002 relating to the assessment and management of environmental noise, OJ L189, of 18 July 2002. Certain categories of noise, such as noise created inside means of transport and noise from domestic activities, are not subject to this Directive.
9. See Murphy E. and King, E.A. (2014), *Environmental noise pollution: Noise mapping, public health, and policy*, Burlington and San Diego, US: Elsevier, p. 4.
10. UN, Economic Commission for Europe and WHO (2012), *THE PEP 2012 Symposium: Green and health-friendly mobility for sustainable urban life*, 21 August 2012 (ECE/AC.21/SC/2012/2 EUDCE1206040/1.9/SC10/2), p. 6.
11. Ibid. p. 3.
12. The Fifth Environmental Action Programme attached to the resolution of 1 February 1993 (15), identifies noise as one of the most pressing environmental problems in urban areas and the need to act with regard to various noise sources. See DO C 138, of 17 May 1993, p. 34.
13. See The Mimi data, The World Hearing Index: Mapping the link between Noise Pollution and Hearing Damage, available at: https://mimi.io/en/hearingindex/, accessed 26 March 2018.
14. See Gray, A. (2017), *These are the cities with the worst noise pollution*, available at: www.weforum.org/agenda/2017/03/these-are-the-cities-with-the-worst-noise-pollution/, accessed 23 January 2018.
15. See EEA, SOER (2015), *The European environment, state and outlook 2015* on Chapter 5, Safeguarding people from environmental risks to heal, available at: https://www.eea.europa.eu/soer-2015/synthesis/report/5-riskstohealth#tab-see-also (last modified 14 January 2016), accessed 23 January 2018.
16. See EEA (2016), *Environmental taxation and EU environmental policies*, EEA Report No. 17/2016, p. 9.
17. Selander, J., Nilsson, M.E., Bluhm, G., Rosenlund, M., Lindqvist, M. and Nise, G. (2009), Long-term exposure to road traffic noise and myocardial infarction, *Epidemiology* 20(2), 272–9. See also the literature review on environmental noise at: http://www.epd.gov.hk/eia/register/report/eiareport/eia_2232014/html/Appendix%20 17.3.2.pdf, accessed 26 March 2018.
18. On the impact on health alone, see OECD (2014), *The cost of air pollution. Heath impacts of road transport*, Paris: OECD Publishing, doi: 10.1787/9789264210448-en.
19. It is extremely important to take into consideration that when the speed of a car exceeds 50 kilometres per hour, the main source of noise is not the engine but the contact of the wheels with the road.
20. Railways are the noise source with the second-highest number of people exposed: 19 million people exposed above 55 dB Lden in the EEA-33.
21. Aircraft noise, with more than 4.1 million people exposed above 55 dB Lden.
22. Industrial noise within urban areas reaches around 1.0 million people exposed.
23. See EEA (2017), *Managing exposure to noise in Europe*, EEA Briefing 01/2017.
24. See Murphy and King (above n. 9), p. 21.
25. See, in particular, *López Ostra v. Spain*, 9 December 1994, Series A no. 303-C (followed by the Spanish Constitutional Court, no. 119/2001, 24 May 2001, *Guerra and Others v. Italy*, of 19 February 1998, but also, regarding noise pollution, *Hatton and Others v. the*

*United Kingdom* [GC], no. 36022/97, *Moreno Gómez v. Spain*, 16 November 2004, no. 4143/02, ECHR 2004-X and *Antonio Martinez Martínez v Spain*, no. 21532/08, §52, 18 October 2011.

26. The first of those cases concerned nuisances caused by a waste-water treatment plant close to the applicant's home, which had affected her daughter's health. The other concerned harmful emissions from a chemical works, which presented serious risks to the applicants, who lived in a nearby municipality.
27. Application no. 55723/00, 9 June 2005.
28. No. 36022/93, 8 July 2003.
29. See paragraph 4. The characterisation of sound as noise is often subjective. Thus, the subjective views of a small minority are acceptable insofar as the emission of that sound does not infringe the fundamental rights of others.
30. See Guidelines for Community Noise – Chapter 4 at http://www.who.int/environ mental_information/Noise/Commnoise4.htm, accessed 26 March 2018; see also Environmental Protection Agency of Ireland at http://www.epa.ie.
31. See, for example, judgment no. 16/2004 of the Spanish Constitutional Court of 23 February 2004, at the legal ground no. 3 and the case-law referred to.
32. It is important to understand this perspective on the development in the EU, focusing the actions not only on the sources of noise but with full understanding of how it is transmitted to the receiver.
33. See Art.1 of the END.
34. The European Commission amended Annex II of the END, in connection with the implementation of CNOSSOS-EU (phase B) in 2012–15. This method must be used to elaborate strategic maps from December 2018.
35. See EEA (2017), Noise fact sheets 2017.
36. Measurements were taken over short periods at strategic locations in the city by mobile noise monitoring terminals and facilitated a prediction model. See http://www. mambiente.munimadrid.es/opencms/opencms/calaire/ContAcustica/SADMAM.html.
37. See European Commission (2012), JRC Reference Reports, CNOSSOS-EU, p. 31, available at: http://publications.jrc.ec.europa.eu/repository/bitstream/111111111/26390/1/cn ossos-eu%20jrc%20reference%20report_final_on%20line%20version_10%20august% 202012.pdf, accessed 26 March 2018.
38. The legal basis is the Federal Act on Railways Noise Abatement (2000) with additional subsequent legislation. See Murphy and King (above n. 9), p. 235.
39. See EEA Report (above n. 16), p. 5.
40. For example, Gago, A., Labandeira, X. and López Otero, X. (2014), A panorama on energy taxes and green tax reforms, *Hacienda Pública Española* 208, 145–90; OECD (2006), *The political economy of environmental related taxes*.
41. See Eurostat (2017), *Environmental tax statistics*. Data extracted in March 2017, available at: http://ec.europa.eu/eurostat/statistics-explained/index.php/Environmental_tax_ statistics#Context.
42. See European Commission (1997), *Environmental taxes and charges in the Single Market*, Brussels, COM (97) 9 final, 26 March 1997.
43. Ibid. p. 20.
44. Ibid. Appendix p. 23.
45. See Alexandre, A., Barde, J-Ph. and Pearce, D.W. (1980), The practical determination of a charge for noise pollution, *Journal of Transport Economics and Policy*, May 1980, p. 219.
46. Alexandre, A. and Barde, J-Ph. (1976), The economics of traffic noise abatement, *Traffic Quarterly*, April, 1976.
47. Alexandre, A., Barde, J-Ph. and Pearce, D.W. (above n. 45), p. 216.
48. See different opinions and arguments at: http://www.sandiegouniontribune.com/busi ness/economy/sd-fi-econometer15jan-20170112-htmlstory.html.
49. Directive 1999/62/EC.
50. See EEA Report (above n. 16), p. 10.
51. Ibid. p. 19.

52. Ibid. pp. 218–19.
53. See Leicester, A. and O'Dea, C. (2008), *The IFS green budget 2008. Aviation taxes*, pp. 187–211. See also, Pearce, B. and Pearce, D. (2000), *Setting environmental taxes for aircraft: a case study of the UK*. Norwich: Centre for Social and Economic Research on the Global Environment.
54. OECD, *Globalisation, Transport and the Environment*, p. 5, available at: www.oecd.org/env/transport/globalisation.
55. For example, in Sweden, Latvia, Hungary, Finland, Bulgaria or France. As some authors point out, indirect taxes on international aviation have some appeal as a prospective source of development financing. See Keen, M. and Strand, J. (2006), *Indirect taxes on international aviation*, IMF Working paper, International Monetary fund, May 2006 (WP/06/124), p. 48.
56. A noise tax would be best targeted to noise pollution (and might be encompassed within airport landing charges). French and Swiss airports currently have such taxes. See http://dbta.dk/wp-content/uploads/2015/10/GBTA-ISSUE-TRACKER-National-aviation-taxes.pdf, accessed 26 March 2018.
57. The new tax on travelling by plane (air passenger duty) was introduced in 1993 and has been increased and restructured more recently.
58. Art. 19 of the amended Law of Finance for 2003 (no 2003-1312 of 30 December 2003) institutes as of 1 January 2005 the 'Tax on Air Transport Noise Pollution' (TNSA). This tax is collected to the benefit of private or public aerodrome administrators for which the annual number of take-offs by aircraft – whose maximum take-off weight is greater than or equal to 20 tons – exceeds 20,000 over one of the previous five calendar years, or, the annual number of aircraft movements – whose maximum take-off weight is at least two metric tons – exceed 50,000 over one of the five previous calendar years, if the 'Noise exposure plan' (PEB) or 'Noise pollution plan' (PGS) of this aerodrome intersects the PEB or PGS of an aerodrome that has all the characteristics previously defined. See https://www.ecologique-solidaire.gouv.fr/en/aeronautical-taxes#e5.
59. There are three groups for the tax rates: (i) 1st group: from EUR 20 to 40 – Aerodromes: Paris-Charles de Gaulle, Paris-Orly, Paris-Le Bourget; (ii) 2nd group: from EUR 10 to 20. Aerodromes: Nantes-Atlantique, Toulouse-Blagnac; and (iii) 3rd group: from EUR 0 to 10. Aerodromes: all the other airports that cross the threshold set in paragraph I of Art. 1609 quatervicies A of the Book of Internal Revenue.
60. See Art.11 of Law 14/2000, of 29 December, which modifies Royal Decree 1064/1991, of 5 July.
61. As stipulated in Art. L.571-14 to L. 571-16 of the Environmental Code.
62. See L.571-16 of the Environmental Code.
63. Standing Committee of the National People's Congress, Order of the President No. 66 of the 12th Congress, Environmental Protection Tax Law of the People's Republic of China. See: http://www.npc.gov.cn/npc/xin,wen/2016-12/25/content_2004993.htm, accessed 23 January 2018.
64. See Income Tax Act, Chap. 134, sections 7 and 19(A) for more details. See also Kuan Liu, H. and Janjua, S. (2009), 'Environmental taxation and its role in Singapore's approach towards environmental sustainable development', in Lin Heng, L. (ed.), *Critical issues in environmental taxation: International and comparative perspectives*, Oxford University Press. Vol. 7.
65. See Tax Incentives in Serbia, Serbian Investment and Export Promotion Agency (www.siepa.sr.gov.yu).
66. Effective: 29 June 2017. History: Amended 2017 Ky. Acts ch. 117, sec. 45, effective 29 June 2017. Created 1974 Ky. Acts ch. 99, sec. 12, effective 21 June 1974. Formerly codified as KRS 224.785.
67. See Kentucky Cabinet for Economic Development: http://www.thinkkentucky.com/kyedc/pdfs/txinpoll.pdf, accessed 26 March 2018.
68. http://www.tax.ohio.gov/sales_and_use/faqs/sales_basics/tabid/3095/QuestionID/433/AFMID/9887/Default.aspx, accessed 26 March 2018.

# 9. Tackling environmental pollution in Seoul, South Korea through tax incentives and related strategies

**Stephanie Lee, Heidi Hylton Meier and Paul J. Lee**

## 1 INTRODUCTION

Many metropolises are suffering from different kinds of pollution. Although it may not seem to be at the forefront of world problems, pollution is a huge detriment to the planet as well as its inhabitants. Scientists have used 'paleoclimates', Earth's historic average global temperature records, to reveal that current climatic warming is occurring much faster than in the past. The two biggest environmental problems of Seoul, South Korea are population and transportation. With over 25 million people, the Seoul Capital Area is considered the second-largest metropolitan population in the world, home to half of all the residents in South Korea.[1] At the end of 2015, over 21 million motor vehicles were registered in South Korea.[2] Many efforts have been made to reduce the increasing pollution and global warming in South Korea but Seoul, and South Korea as a whole, are far from avoiding the serious detriments of pollution it has already created.

## 2 AIR POLLUTION

The air quality of Seoul has been improving in recent years due to the efforts of the city. The amount of greenhouse gas production of South Korea is ranked at number seven compared to other OECD countries. The rate of emissions per capita, however, is ranked number one. By 2030, South Korea has pledged to reduce its emissions by 37 percent.[3] However, there are still many causes of poor air quality in South Korea. Every April, 'Hwang Sa', also known as yellow dust, flies from the deserts of China and Mongolia causing problems for the people living in Korea to partake in everyday life. This yellow dust not only reduces visibility for about a month, but also contains viruses, fungi, bacteria, industrial pollutants (such as pesticides) and heavy metals. These pollutants in the yellow dust harm not only humans, but also crops and soil. There is not only physical damage from yellow dust, but also economic drawbacks. Outdoor vendors are forced to close their shops, outdoor equipment and inventory can be damaged, and decreased foot traffic (but increased auto transportation) for store owners are just some of the immediate ramifications of the yellow dust. The National Emergency Management Agency (NEMA) states that both the agricultural and livestock industries have to disinfect and clean barns, pastures, and greenhouses in order to avoid animals contracting dust-borne diseases.[4]

Another source of poor air quality is from coal-fired power plants located in South Korea. Ultra-fine dust is emitted into the air from power production which contains sulfur oxides and nitrogen. These particles can invade the lungs and the bloodstream. The nitrogen and sulfur oxides found in the ultra-fine dust, along with the cadmium, lead, and arsenic found in the yellow dust, cause short- and long-term damage to South Korea's inhabitants and cause around 50 percent of Seoul's pollution.[5]

Reports claim that over half of the smog caused by ultra-fine dust and yellow dust is the fault of China. There are approximately 1,600 deaths each year due to the dust and this number is expected to increase to 2,800 per year if pollution in South Korea is not constrained.[6] The new president of South Korea has proposed ten initiatives in order to reduce fine dust and to solve air pollution problems.[7] Fourteen percent of the fine dust is from coal thermal power plants. There are currently 59 coal thermal power plants in South Korea, with ten plants being over 30 years old. Compared to the newer plants, more than twice the amount of the fine dust is produced from the older plants than from the newer ones.[8] Six days after the new president was elected, the ten older power plants were temporarily shut down. The new president has mentioned he would like to shut down all the coal thermal power plants during his term in office. The new president has also promised to install air purification systems in all indoor school gymnasiums. For the government, he has set up a task force department to monitor fine dust emissions.[9]

Seoul has also taken measures to implement solutions to decrease air pollution. Climate Action Tracker states that 'South Korea intends to achieve part of this air pollution reduction by using "carbon credits from international market mechanisms".'[10] The South Korean government has stated that 'a 25.7 percent reduction in air pollution will be achieved domestically and a further 11.3 percent reduction will be achieved by international market mechanisms'.[11]

Seoul has been improving public transportation to lessen pollution. Beginning in 2000, 8,750 buses in Seoul were replaced with buses that use natural gas. Although this policy cost the city $350,000,000, it benefited the residents of Seoul due to lower emissions of fine dust and three times less emissions of nitrogen oxide.

Old diesel cars are also a problem in South Korea. The government has given a grant of approximately $7,000 to owners of old diesel cars as an incentive to scrap the car. If the diesel car is more than seven years old, the government gives a grant of approximately $15,000 if owners scrap the car early. The owner of the diesel car can sell the car as well if they take it to the junk yard. In 2017, the government gave grant support of $10,500 for scrapping diesel cars early. If most owners do choose to scrap their diesel cars, the government is expecting to reduce emissions of nitrogen oxide and fine dust by 1,172 tons. Compared to new cars, old diesel cars emit 5.8 percent more pollutants than the newer cars.

Seoul also encourages people to use bicycles instead of driving a car. This will decrease car traffic along with air pollution, while also implementing physical exercise for those who cycle and walk. This policy was introduced in 2007, but despite its efforts, the cycling population has not increased substantially. Another solution to benefit the environment of South Korea is to increase the number of parks. Trees and plants filter pollutants and dust from the air, providing cleaner air to breathe and lowering temperatures in urban areas. The plant life in these parks will reduce erosion while also cultivating a more vibrant ecosystem to host small animals.

## 3 FOOD WASTE POLLUTION

Approximately 1.3 billion tons of food is wasted annually in South Korea, with food being thrown away or rotting in the fields. It is not just the financial cost of paying for the food and not consuming it, but also the economic and environmental cost of disposal. Not only is the output of food waste in South Korea extremely high, food waste is extremely difficult to manage. Seoul has been a pilot city to implement special trash bins that weigh and fine those who discard too much food (Food Waste Balance System). Since 2010, Seoul has had a grant plan to lower food waste. Currently, 17,000 tons of food waste is produced daily but a goal has been set to reduce the waste to 14,000 tons. The Food Waste Balance System started in June 2013 and has been implemented everywhere in Seoul. Fines and regulations have been put into place since the Food Waste Balance System was enacted. People who produce food waste are charged (for example, restaurants, etc) and taxed for producing food waste as well as being charged for collection and disposal costs. Seoul has eliminated free collection and disposal of waste. Since the Food Waste Balance System was implemented, the amount of food waste has decreased every year.

Seoul took a poll about the Food Waste Balance System and the government's education on food waste. In 2010, 68.8 percent of respondents thought there needed to be a food waste system. In 2017, 87.6 percent thought Seoul needed a food waste system and should continue to implement it.[12] From apartment buildings to giant hotel kitchens, leftover food in Seoul is picked up and taken to sorting facilities where it is crushed and dried for animal feed or fertilizer or burned to generate electricity. Trial districts in Seoul have succeeded in reducing household food waste by 30 percent and restaurant food waste by 40 percent. Programs are now underway in 90 localities in South Korea. The goal is not only to drastically curtail food waste, but also to process or incinerate all of South Korea's remaining leftover food, thus keeping it out of landfills where it will decay and emit methane, a potent greenhouse gas.[13]

Not only does food waste contribute to greenhouse gas emissions, certain foods create an even higher level of emissions than other foods. Although the Industrialized Asia region, which consists of China, Japan and South Korea, is a leader in food waste, it comes in second to North America and Oceania when it comes to food waste per capita.[14] Figures 9.1, 9.2, and 9.3 show the contribution of each commodity to food wastage and carbon footprint, the contribution of each region to food wastage and carbon footprint, and the carbon footprint of food wastage by region respectively.

## 4 WATER POLLUTION

There are many factors that contribute to water pollution. Some of the causes are littering, improper waste management, detergents, and lack of care. In order to lessen water pollution and reduce 6.9 billion tons of water waste per year, Seoul has mandated water pipe flushing.[15] Currently, the satisfaction rate with drinking water is 59 percent for the water supply. Seoul has set a standard of 80 percent satisfaction. Four goals were set: create a water pipe system and department to manage and oversee the system; clean and create safer drinking water; increase drinking water quality; and make information public for consumers.[16] Old water pipe systems are to be modernized, smart sensor Internet of Things (IOT) added, and new sensors and improved technology are to be used.

Polluted bodies of water can destroy ecosystems and contaminate groundwater. South Korea's water has a history of heavy pollution. There

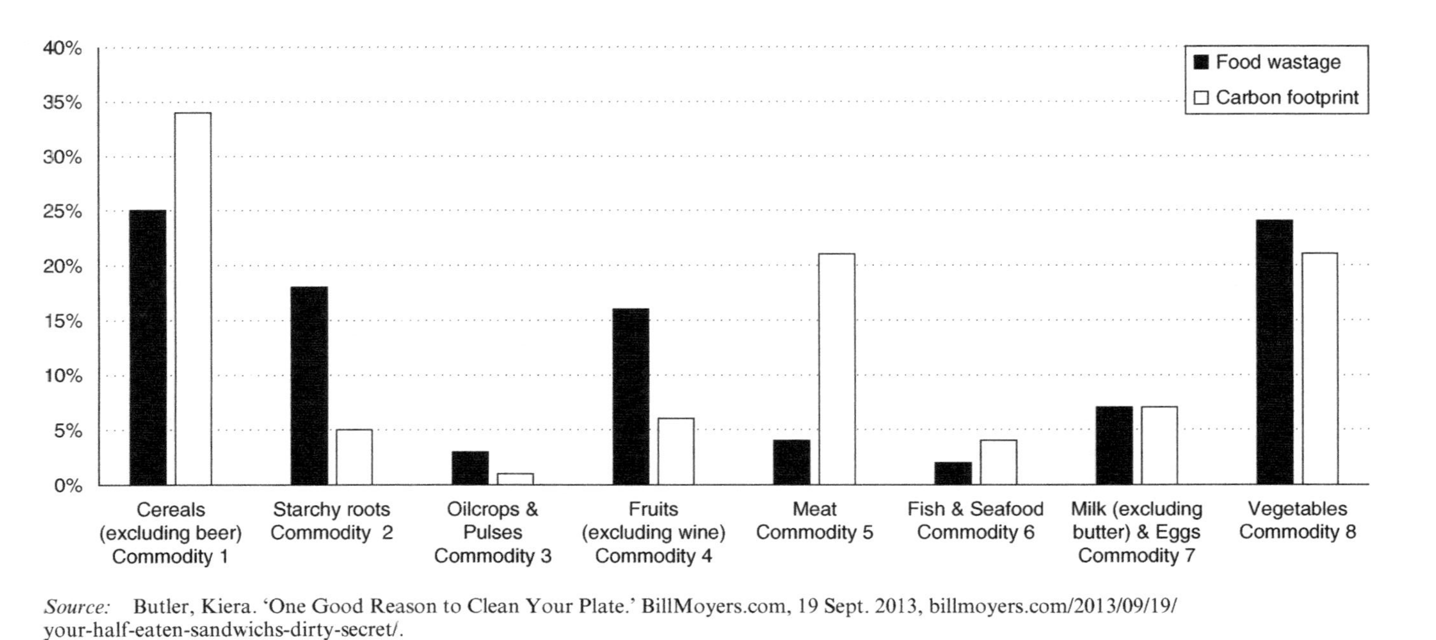


*Source:* Butler, Kiera. 'One Good Reason to Clean Your Plate.' BillMoyers.com, 19 Sept. 2013, billmoyers.com/2013/09/19/your-half-eaten-sandwichs-dirty-secret/.

*Figure 9.1 Contribution of each commodity to food wastage and carbon footprint*

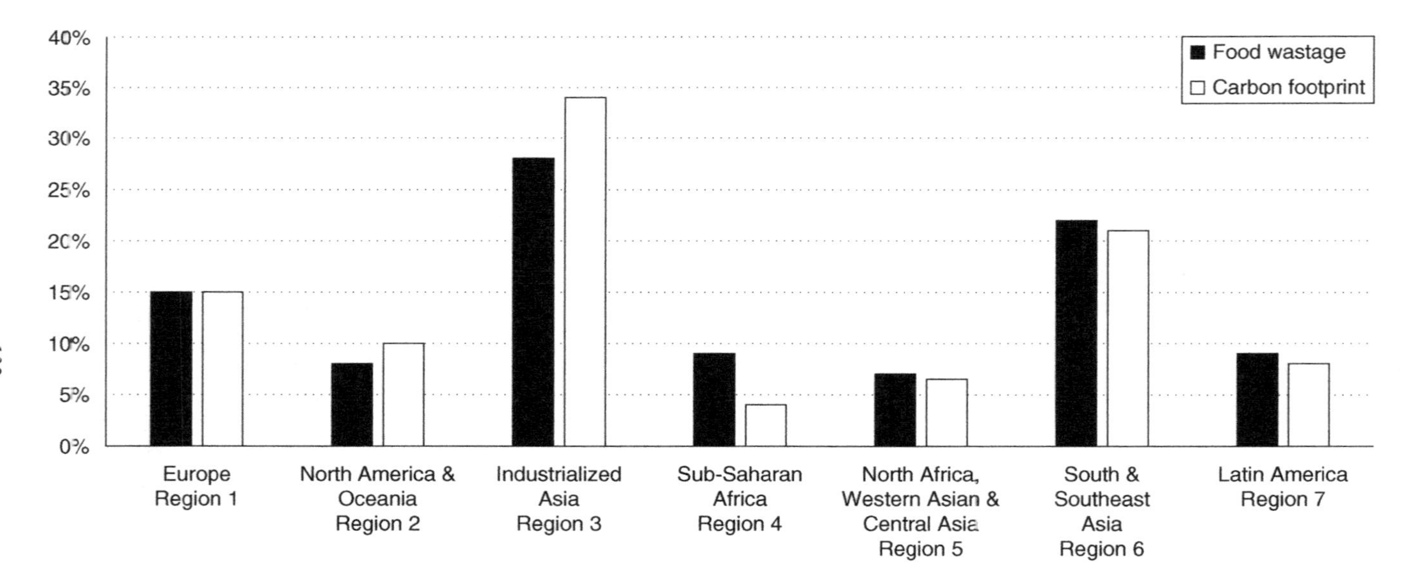


*Source:* Butler, Kiera. 'One Good Reason to Clean Your Plate.' BillMoyers.com, 19 Sept. 2013, billmoyers.com/2013/09/19/your-half-eaten-sandwichs-dirty-secret/.

*Figure 9.2 Contribution of each region to food wastage and carbon footprint*

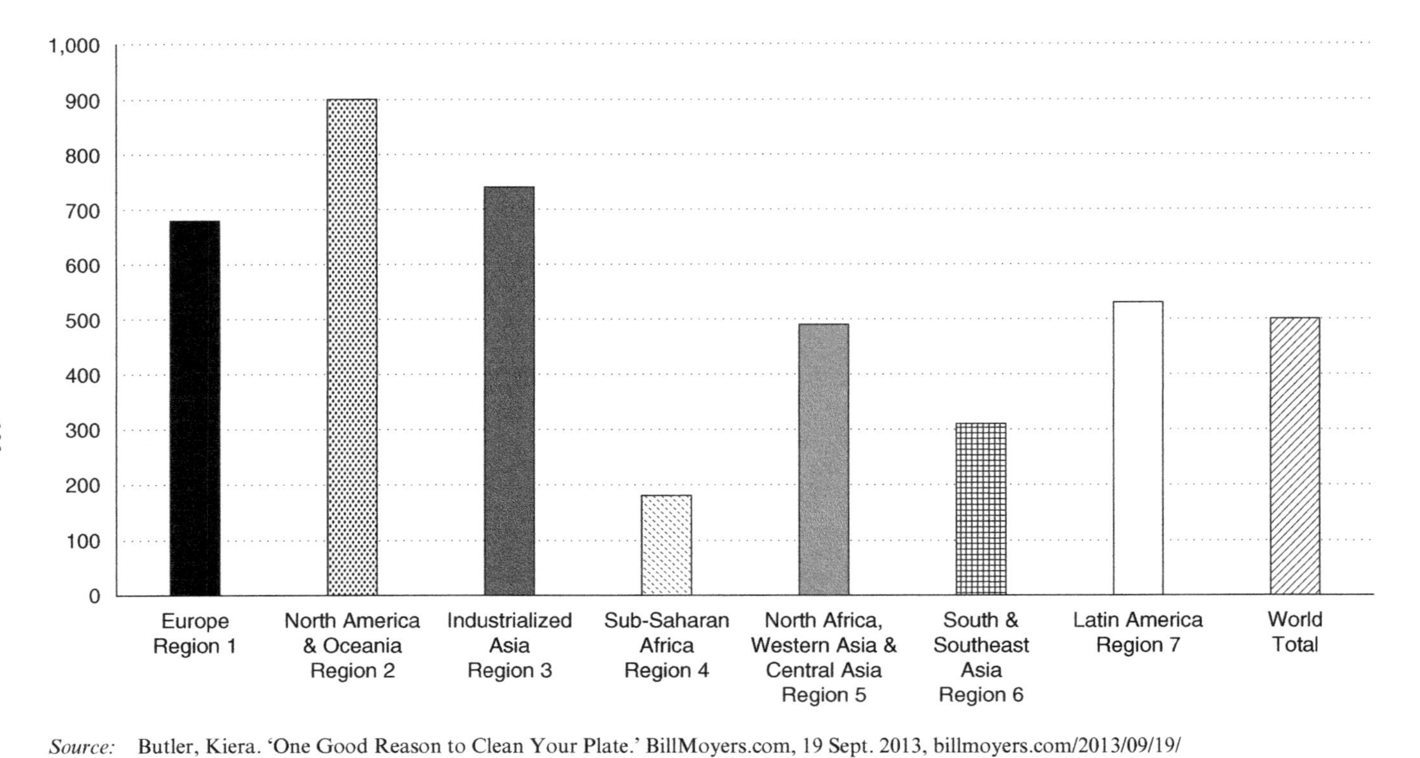


*Source:* Butler, Kiera. 'One Good Reason to Clean Your Plate.' BillMoyers.com, 19 Sept. 2013, billmoyers.com/2013/09/19/your-half-eaten-sandwichs-dirty-secret/.

*Figure 9.3 Carbon footprint of food wastage, by region*

have been many projects to restore the water quality of rivers, which have led to some reduction in water pollution. In 2008, the Four Major Rivers Restoration Project was started as part of South Korea's 'Green New Deal' policy. This $18 billion project aims to restore the water quality of South Korea's four biggest rivers – the Han, Nakdong, Geum and Yeongsan, along with hundreds of miles of 14 tributary streams. Projects such as rainwater harvesting are one of many initiatives to secure water resources to safeguard against possible water scarcity, and for flood control, water quality improvement, and revitalizing river ecosystems. The three phases of the Four Major Rivers Restoration Project, two of which are already complete, have improved water quality. Phase one was the complete restoration of the four major rivers, while phase two restored all of the major tributaries. Phase three involves revitalization of about 929 km of additional tributaries.[17]

The three phases of the Four Major Rivers Restoration Project involved dredging the rivers and installing two barriers in the river, a fixed barrier to manage the water levels and a moveable barrier to prevent flooding. These small dams have helped with flood control and water management, while existing agricultural reservoirs have been making the banks higher. Jae Park, University of Wisconsin-Madison, professor of civil and environmental engineering, states:

> I worry that Korea will have more severe drought and flooding cycles and will need more water resources and water management projects. The Four Rivers Restoration Project is just the beginning. In the last 40 years, investment in rivers has been fairly minimal compared to road and rail. Korea now has a very good road and rail system, so it is a timely project (to improve the rivers).[18]

## 5 SOLUTIONS TO ENVIRONMENTAL PROBLEMS

Many projects have been implemented to decrease pollution in South Korea. Whether it is transportation, water cleaning, or food waste management, pollution levels have been improved. South Korea has increased the acreage of green space, as well as investing in cycle paths that are readily available for the people to use on a regular basis. From recycling to waste management, these projects have been vital in protecting, restoring and creating a revitalized environment for South Korea.

Since 2013, South Korea has made it a top priority to reduce greenhouse gas emissions and 'plans to launch the world's most ambitious carbon-trading market to cut emission levels by 30 percent by 2020.'[19] South Korea will attempt to reduce greenhouse emissions while avoiding harm to the family-controlled conglomerates that supply the world with technology and products. Bloomberg New Energy Finance analyst Richard Chatterton stated that this ambition of South Korea will be extremely costly and highly burdensome compared to other countries.[20] This is due to South Korea closing the loopholes that caused carbon price declines in the EU, which allowed polluting industries to not reduce the amount of their emissions.

> For one thing, South Korea plans to restrict the use of so-called offsets. An offset is essentially a license to pollute by paying someone else to reduce emissions on your behalf. Industrial companies can buy offsets from international organizations that run climate-friendly projects such as planting trees or capturing methane emissions from landfills. South Korean companies, however, will be able to purchase carbon credits only from domestic projects and only until 2021. After 2021, companies can purchase international carbon credits but offsets can only be used to meet no more than 10 percent of emissions-reduction targets.[21]

Nonetheless, this will cause South Korean goods to be more expensive, which could slow the economy because companies will have to pay expensive emissions allowances or implement extensive measures to reduce carbon pollution.[22]

Sungwoo Kim, regional head of climate change and sustainability at KPMG in Seoul, states that there will be a miniscule impact on overseas Korean smartphone buyers because responsibility for emission payments will fall on steel makers and electricity companies. 'It will be easier for electronics companies, like Samsung, to bring their emissions down by capturing fluorinated gases that are emitted in the production of semiconductors and video screens.'[23] Although many South Korea companies are worried about the expense of a carbon market, the government and conglomerates are looking at it optimistically as a new opportunity for economic growth.

'Politically, South Korea has shown that it is eager to be seen as a leader in the development of global climate change policy, so it has taken on a strong target to justify this.'[24] Figure 9.4 shows a comparison between different emission trading schemes between 2015 and 2020. Some of the biggest emitters are power companies that are government owned.

Since the initiation of the cap-and-trade system in 2015, South Korea has become the world's second-largest carbon trading market and also aims to reduce pollution in other ways since it is the seventh-largest annual greenhouse gas emitter after China, the US, India, Russia, Japan, and Germany.

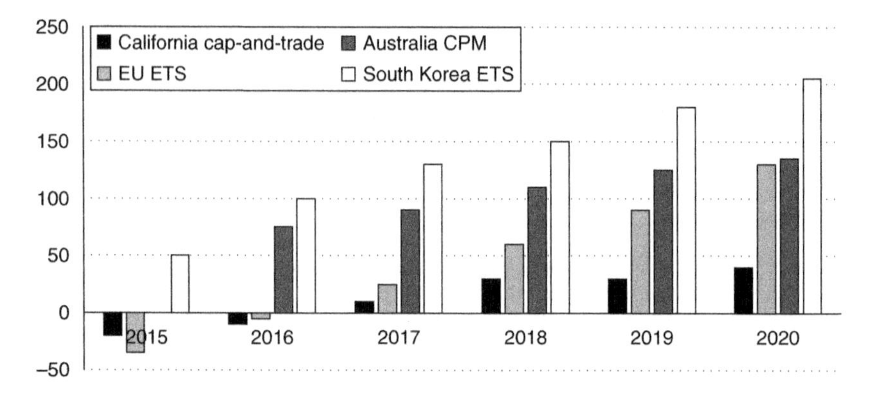


*Note:* Excludes historically banked volumes of allowances.

*Source:* Bloomberg New Energy Finance.

*Figure 9.4 Estimated annual demand for abatement excluding usage of offsets, comparison between different emission trading schemes, 2015–20 (Mt$CO_2$e)*

> The country is hoping to cut its transportation emissions by about 34 percent, its power generation emissions by about 26 percent and its public sector emissions by about 25 percent. In addition to the cap-and-trade system, South Korea also plans to create an 'Eco-friendly Transportation Campaign' and campaigns calling on South Koreans to dress according to the season in order to save energy spent on heating and cooling. Some cities in South Korea have also taken steps to curb climate change, In October 2013, residents of the city of Suwon parked their cars for 30 days, opting for more climate-friendly transportation options such as walking or biking.[25]

The cap-and-trade market of South Korea includes 525 of the country's largest polluters, which account for approximately two-thirds of the nation's non-vehicular emissions. These include power generators, petrochemical firms, steel producers, car makers, airlines and electromechanical firms.[26] With a fixed number of permits to cover its own emissions, each entity must watch emissions to make sure their allocated permits cover their emission allotment, or anticipate incurring heavy fines.

Since the effort to protect and clean the environment in South Korea started, there have been many different policies implemented. These environmental policies, however, have been insufficient to protect or preserve the environment and its resources. The Yale Environmental Performance Index from 2014 shows 'Korea ranked at 43rd out of 178 countries in overall environmental quality and ranked at 166th place in terms of air quality.'[27]

## 6 CONCLUSIONS

There have been many suggested solutions to reduce pollution in Seoul and South Korea. With these in mind, the next 15 years will be important for the environment in South Korea. If participants follow the new cap-and-trade system, the environment of South Korea will improve drastically while also setting a standard for other countries to follow. However, if not done well, the system could be a waste of time, money, and resources. Although it is quite ambitious for South Korea to project emission cuts by 30 percent in 15 years, if achieved, South Korea will be an exemplary nation that can help other nations implement their own cap-and-trade system in the most efficient way. If South Korea is successful in its efforts, the country will take a tremendous leap forward in improving its environment.

## NOTES

1. 'Seoul Population 2017.' Seoul Population 2017 – World Population Review, world-populationreview.com/world-cities/seoul-population/.
2. Choi, Sung-jin. 'Number of Cars in Korea Nears 21 Million.' Korea Times, 17 Jan. 2016, www.koreatimes.co.kr/www/news/biz/2016/01/123_195662.html.
3. Kim, Yvonne. 'Hwang Sa (Yellow Dust).' Asia Society, 14 Feb. 2014, asiasociety.org/korea/hwang-sa-yellow-dust.
4. ibid.
5. http://www.upi.com/Top_News/World-News/2015/03/23/Pollution-in-South-Korea-poses-increasing-health-threats/5191427123633/.
6. http://www.ibtimes.com/smog-korea-coal-power-plants-not-china-are-major-culprits-says-greenpeace-1835638.
7. Myoung, Min Jun. 'Airpocolypse – Unintentional Fine Dust, until When?' Sbs, 17 May 2017, sbscnbc.sbs.co.kr/read.jsp?pmArticleId=10000859196&pc_searchclick=sub_news_total_01_01.
8. ibid.
9. ibid.
10. Climate Action Tracker. 'South Korea.' CLIMATE ACTION TRACKER, climateactiontracker.org/countries/southkorea.html.
11. 'Environmental Policy in South Korea.' Environmental Policy in South Korea, 2015, www.kas.de/wf/doc/17802-1442-1-30.pdf.
12. 'South Korea's Food Waste Solution: You Waste, You Pay.' ASIA TODAY, www.asiatoday.com/pressrelease/south-koreas-food-waste-solution-you-waste-you-pay.
13. Chrobog, Karim. 'In South Korea, An Innovative Push to Cut Back on Food Waste.' Yale E360, 20 May 2015, e360.yale.edu/feature/in_south_korea_an_innovative_push_to_cut_back_on_food_waste/2875/.
14. Butler, Kiera. 'One Good Reason to Clean Your Plate.' BillMoyers.com, 19 Sept. 2013, billmoyers.com/2013/09/19/your-half-eaten-sandwichs-dirty-secret/.
15. Choi, Young-Jun. 'Water Management Policy of the City of Seoul.' Seoul Solution, 9 Feb. 2017, www.seoulsolution.kr/en/content/water-management-policy-city-seoul.
16. ibid.
17. 'South Korea's Four Rivers Restoration.' Water and Wastewater International, www.

waterworld.com/articles/wwi/print/volume-27/issue-6/regional-spotlight-asia-pacific/south-korea-s-four-rivers-restoration.html.

18. 'South Korea's Four Rivers Restoration.' Water and Wastewater International, www.waterworld.com/articles/wwi/print/volume-27/issue-6/regional-spotlight-asia-pacific/south-korea-s-four-rivers-restoration.html.
19. Woody, Todd. 'South Korea Goes It Alone with the World's Most Aggressive Carbon Market.' Quartz. N.p., 23 May 2013. <http://qz.com/87452/south-korea-goes-it-alone-with-the-worlds-most-aggressive-carbon-market/>.
20. 'South Korea's Emissions Trading Scheme, Bloomberg New Energy Finance.' Bloomberg New Energy Finance. N.p., 09 May 2013. <http://about.bnef.com/white-papers/south-koreas-emissions-trading-scheme/>.
21. Woody, Todd. 'South Korea Goes It Alone with the World's Most Aggressive Carbon Market.' Quartz. N.p., 23 May 2013. <http://qz.com/87452/south-korea-goes-it-alone-with-the-worlds-most-aggressive-carbon-market/>.
22. ibid.
23. ibid.
24. ibid.
25. Valentine, Katie. 'South Korea Launches World's Second-Largest Carbon Trading Market.' ThinkProgress RSS. N.p., 12 Jan. 2015. <http://thinkprogress.org/climate/2015/01/12/3610553/south-korea-cap-and-trade/>.
26. Dechert, Sandy. 'New South Korea Cap-And-Trade Market Becomes World's Second-Largest.' CleanTechnica. N.p., 14 Jan. 2015. <http://cleantechnica.com/2015/01/14/new-south-korea-cap-and-trade-market/>.
27. Stiftung, Bertelsmann. 'South Korea.' SGI 2016, South Korea Environmental Policies, www.sgi-network.org/2016/South_Korea/Environmental_Policies.

# PART III

# Revenue perspectives

# 10. Green ICMS: Brazil's tax revenue distribution based on environmental criteria*

**Lise Tupiassu, Bernardo Mendonça Nobrega and Jean-Raphaël Gros-Désormaux**

## 1 INTRODUCTION

Brazil is a country of continental proportions and has in its territory exuberant fauna and flora. It is also an extremely unequal country, both socially and economically, organized in a federative way. Mixing these elements creates a unique challenge: how to develop the country and promote greater social and economic equity, while preserving its environmental wealth.

Despite heterogeneity being a rule in federative models, in Brazil this difference is so exacerbated that, both internally and intra-federatively, the situation is quite drastic (Rubinstein, 2010). This makes the duality of economic development versus environmental protection much more pronounced, which makes it difficult to find a break-even point.

Environmental taxes and other economic instruments are commonly used to integrate economic development and environmental purposes. Beyond imposing an onus on the polluter to promote environmental internalization (Marshall, 1920; Pigou, 1932), environmental taxation generates income for the State that can be used either within a strict ecological perspective or in a wider sphere for social reorganization of the State.

In Brazil, those instruments are not used in a systematic way. The country has yet to develop specific environmental taxes, but sometimes uses the extrabudgetary consequences of its current fiscal institutions to achieve environmental protection (Domingues, 2012). Several solutions are proposed and attempted. One of the most successful experiences is the economic-tax instrument called the Green ICMS (or Ecological ICMS). It is a fiscal federalism mechanism that allows intergovernmental fiscal transfers based on environmental criteria.

The objective of this chapter is to present this Brazilian experience and the challenges faced in the application and use of this instrument. It aims to show how the instrument works, its achievements and its challenges.

## 2 WHAT IS THE GREEN ICMS?

To first understand the Green ICMS the basic notion of the ICMS and Brazil's fiscal federalism must be explained. The primary tax legislation, including the stipulation of all taxes used by Brazil's federal, State and counties governments, is established in the Federal Constitution. The State's main tax is called, in Portuguese, *Imposto sobre a Circulação de Mercadorias e Serviços* – ICMS – which translates as Service and Merchandise Circulation Tax. It is, in summary, a sales tax or a value-added tax. It is a very important source of revenue for local governments.

Fiscal federalism in Brazil's Constitution determines that of all the revenue the State collects from the ICMS, 25 percent must be passed to the counties. 18.75 percent of the value is distributed based on the amount of value added to goods in the county; and 6.25 percent is transferred as freely determined by State law.

This distribution system traditionally favors the most economically developed municipalities, since most of the ICMS revenue transferred to the counties is based on the amount of sales tax generated in each one. Normally, the more developed cities are the ones most capable of generating a high tax revenue from the circulation of goods.

The counties that have environmental conservation areas, water reservoirs and indigenous lands request to change it, because they were submitted to double penalties: on the one hand, they were confronted with restrictions on the economically productive use of part of their territory due to the environmental allocation. On the other hand, this restriction had economically disastrous consequences, which implied a reduction in the level of budgetary revenues.

Fernando Scaff and Lise Tupiassu (2005, p. 735) explain that the way ICMS distribution works is detrimental to the environment because it rewards those that invest more in economic development to the detriment of environmental protection. Also, the counties that preserve nature and the positive externalities end up with a smaller amount of ICMS funds. That means that in the traditional system, those who generate positive externalities are penalized. In fact, the logic of distribution of ICMS revenues was detrimental to counties that, with land restrictions, produced positive environmental externalities for all others.

In addition, it is important to notice two important things about the land restrictions. First, they make up 37.1 percent of the Brazilian territory

(Embrapa, 2017), and second, they are concentrated mostly in Brazil's north, a region that has low economic and social development in comparison to the rest of the country.[1]

In fact, the federal system of public protected areas establishes a large number of conservation units (Drummond et al, 2009), and almost 40 percent of them are located in Brazil's northern region (Drummond et al, 2012). States and counties also create environmentally protected areas, but all those land use restrictions were considered as 'preventing them from developing productive activities and generating value added' (Grieg-Gran, 2000, p. 1).

The counties claimed a fair distribution in the ICMS intergovernmental transfers to compensate for those environmental land restrictions.[2] Finally, as regards the 6.25 percent of ICMS revenue for which the Constitution gave greater freedom of distribution, the State began to condition the transfer of this part on the fulfillment of socio-environmental requirements in a system that was called the Green ICMS (or Ecological ICMS). That means that a portion of the ICMS revenue transferred to counties started to be determined by environmental conditions, inserting some ecological criteria in the Brazilian fiscal federalism (Tupiassu, 2006).

Thus, the Green ICMS was created as a sales tax revenue distribution instrument that aims to fight disparities in Brazil's fiscal federation model, which privileges municipalities with higher economic development, conditioning the distribution of said funds on environmental criteria. It focuses on municipalities that have land restrictions that hinder economic development and those that have not yet adopted a predatory economic model giving them an alternative income.

This provision of financial compensation for non-degradation is one of several ways of using economic instruments for environmental purposes, serving as a means to meet the demand for sustainable development resources and, above all, preservationist policies.

The use of the Ecological ICMS clearly applies not only the polluter-pays principle but also its closest corollary, represented by the protector-beneficiary principle, in light of the fact that those who pollute receive nothing – on the contrary, lose – while those who preserve are rewarded.

As defended by Rosembuj (1995) it is a logical and necessary corollary of the polluter-pays principle that all those who create, by their conduct, specific situations of environmental conservation, benefiting all, should receive fair compensation, with due acknowledgment of the positive externalities of those whose environmental behavior reduces public spending and brings benefits to the whole community.

Based on all these principles, the Ecological ICMS was created in the

State of Paraná in 1991. Soon, the policy spread to other Brazilian States, and the number of municipalities benefited by the institute increases every day, mostly because of the very positive effects of this policy.

## 3 POSITIVE EFFECTS OF THE GREEN ICMS

Traditional environmental taxation follows a Pigouvian (Pigou, 1932) approach, where negative externalities that are not internalized by the market are internalized by a tax focused on the negative externality-producing activities.[3]

Even if some fiscal instruments are often called environmental taxes,[4] Brazil does not have a generalized tax policy focused on environmental issues. In fact, due to Brazil's rigid fiscal system, an environment-focused tax cannot be easily created, which is why José Marcos Domingues (2012, p. 51), explains that the main instruments for pollution control in Brazil are not fiscal instruments:

> Brazil has much to learn from more developed countries when it comes to establishing strict-sense environmental imposts. So far, there has been no such specific tax on carbon, or on emissions or on effluents. Like Japan's national government, what the Brazilian authorities mostly do in this area is to apply command and control mechanisms, in other words, increasingly demanding licensing and environmental pattern.

Facing this reality, the Green ICMS consists of a very interesting environmental protection mechanism. Unlike several environmental taxation instruments, it does not involve the creation of a new tax. After all, it does not present any burden to the taxpayer, only the distribution of resources collected by an existing tax.

As an instrument based on the existing system, the Green ICMS is relatively easy to introduce because it needs only a change in the State's law. It affects only the revenue transferred by State to cities to compensate environmental efforts. The logic behind this reasoning is quite simple: If one considers the premise that economic development is attained through land use as the main instrument, those that cannot make use of it are then compensated. It is a financial possibility for municipalities that cannot generate enough sales tax to guarantee a high income.

The fiscal instrument aims at giving priority to those municipalities with less economic development and which, consequently, receive a lower share of ICMS resources. However, the compensatory intention marking the beginning of this policy has been replaced by an incentive effect as a growing number of cities have begun to implement environmental policies through the creation of natural parks for example, which are eager to receive part of the financial resources granted on ecological criteria.

The positive effects of this policy are very clear in the State of Paraná. Since the beginning there has been an enlarged number of municipalities increasing their total protected areas aiming to receive more revenue from the Ecological ICMS transfer, as demonstrated by Young and Roncisvalle (2002, p. 33):

> This tax redistribution system has been very effective in encouraging the *municipios* to increase the total protected area in their boundaries, since this would represent a higher budget. For example, the *municipio* of Morretes was the 203rd in the ranking of tax redistribution before 1992, and after the law it became the 107th, while the *municipio* of Antonina moved from the 191st to the 84th position. Another indication that the law has been successful is that the number of *municipios* that are considered eligible for the benefit increased from 112 in 1992 to 192 in 1998.

The expectation is that, following the logic proposed, municipalities with larger areas of conservation and greater investments in environmental management will be rewarded with a higher transfer of revenues. In this way, in order to obtain a higher amount of resources, the municipalities will always try to invest in the preservation and improvement of environmental conditions, initiating a beneficial fiscal war.[5] This is what Loureiro (2008, p. 14) calls a 'virtuous circle'.

The 'virtuous circle' notion is that cities that fulfill the environmental criteria have increased financial power, which in turn helps those municipalities improve their ability to protect the environment and fulfill the Green ICMS's criteria to an even greater extent. This way, the better the environmental protection, the more the municipality wins and the more environmental protection can be obtained.

In addition to these benefits, the Ecological ICMS has been contributing to the development of new mechanisms to protect the environment, including environmental education projects, ecological orientation of rural producers, improvement of the quality of the water, implementation of policies for solid waste, mass investment in indigenous territories and the creation of natural heritage reserves,[6] among many others.

Some municipalities have also adopted social criteria, benefiting not only the municipalities that have conservation units, but also those that have a sewage treatment system or adequate garbage disposal, and introducing criteria involving education, historical patrimony and health, among others. This means that not only environmental protection would be compensated, but also the living conditions of the population would be improved. Considering Brazil's high poverty rates, with human development among the lowest in the world, it is a real revolution in the form of encouraging the construction of public policies.

The experience was considered a success and from its implementation in the State of Minas Gerais,[7] which brought surprising redistributive results, became widespread. In the first year – when the definitive indexes were not yet in place – almost 500 municipalities obtained revenue increases of more than 100 percent, and in 38 of them, the increase exceeded 1,000 percent. The minimum per capita portion, which was R$[8] 0.88, increased to R$ 15.12, while the maximum per capita portion of R$ 684.53 decreased to R$ 587.99 (Riani, 1997).

## 4 DEFINING ENVIRONMENTAL CRITERIA FOR INTERGOVERNMENTAL SHARING

Due to larger positive effects, the Green ICMS is now used by 18 of the 26 Brazilian States, each one of them having the liberty to establish the amount and the environmental criteria to use in its own transfers. Nevertheless, the choice of criteria is the most critical decision faced by the policy makers. The criteria must ideally correct a fiscal distortion while providing effective environmental protection. Without fulfilling these two requirements, the goals of the policy are greatly compromised.

In general, the criteria adopted by the States are based on quantitative or qualitative issues. The quantitative criteria are adopted by the legislation of all States holding the Ecological ICMS. These criteria are purely objective in nature. They define the amount of Green ICMS revenue according to the total protected area, or its ecological and social importance. An example of a quantitative criterion is the size of conservation units or environmentally protected areas compared to the total area in a county's territory.[9]

This type of criterion does not make any further or profound analysis of the environmental quality inside the conservation unit, nor does it care about its actual implementation or current state. There is no practical subjective focus. The question in this analysis is purely objective, so that only the percentage of the territory occupied by Protected and Special Use Areas matters, regardless of the conservation state or protection power of these areas.

These kind of criteria makes the policy implementation easier, since they demand a sort of geographic information commonly accessible, and do not require complex calculations to establish the municipalities' share. However, the use of the quantitative criteria as the only variable to determine the distribution of revenues between municipalities can present serious distortions.

The Amazonian State of Pará has many examples of this distortion. Studies (IDESP, 2013; Oliveira, 2014; Oliveira and Tupiassu, 2016) identified that, in this State, some of the municipalities that received the highest quotas of Green ICMS funds were the ones that presented the highest deforestation levels. Altamira, Senador José Porfírio, São Felix do Xingu, Anapu and Ourilandia do Norte, some of the major receivers of the Green ICMS, are all included on the Ministry of the Environment list of those with the biggest deforestation areas in Brazil.

This problem occurs because those municipalities have very large conservation units and very important indigenous lands. Nevertheless, those territories, normally considered as the best way to protect the Amazonian forest (Nolte et al, 2013; Nunes, 2010), are being devastated to a large extent. This shows that when the policy considers only quantitative criteria, the income distribution does not reflect the real environmental condition in those areas.

Few States, in turn, use qualitative criteria. They are of a subjective nature, since they are linked to analysis of the environmental quality and not only its existence. That is, they aim to make a more in-depth and subjective analysis of certain criteria. An example of a qualitative criterion is the analysis of the state of preservation of a given conservation unit, verifying the actual conservation of the biodiversity or the quality of the water resources of a given area.

Rio de Janeiro legislation, for instance, establishes, as one of the criteria for the distribution of the Green ICMS, the environmental quality index of water resources.[10] The State of Tocantins also stipulates that conservation units and indigenous lands will be analyzed following qualitative criteria proposed by the Nature Institute of Tocantins and approved by the State Environmental Council.[11]

In this way, we can see that the qualitative criteria bring special attention to the Green ICMS since they increase the concern with effective environmental protection, conditioning the transfer of funds on criteria that require practical results. These kind of criteria, consequently, require a heavy monitoring scheme in order to control environmental quality standards, which is not easy to implement in practice in all States. As Seroa da Mota (2000) explains:

> the criterion for performance indicators needs to be well defined and the EPAs have to be institutionally strong to properly apply them. Without the proper application of performance indicators, incentives for conservation will fail as beneficiaries will not perceive changes in their shares to reflect changes in the degree of land degradation. In fact, without adequate monitoring, this instrument may instead be a disincentive to conservation.

Choosing a criterion is not an easy task. Due to the enormous difference between States, the sharing of experiences and information is, to a certain extent, hard. It is very complicated to compare a State in the Amazon, with unique challenges, with others further to the south. Furthermore, the institute is relatively new in several States and there is still insufficient information on the criteria used to define with certainty which ones are the best.

The legislation of the State of Pará, in a wise decision, established the obligation to review the adopted criteria. Article 6 of State Decree No 775 of 2013 determined that the proposed criteria and its indicators should be reassessed in the year 2015. The legislator, in this case, foresaw the possibility that the chosen criteria might not have been best and established a period of application of two years, with ulterior evaluation of the results as a means to allow the institute to evolve.

This period of evaluation adopted in the legislation of the State of Pará can be considered very important. This is because it allows the correction of a possible distortion caused by the application of those criteria that offend the two main objectives of the institute.

## 5 CHALLENGES IN THE GREEN ICMS IMPLEMENTATION

Ecological ICMS policy represents a clear and positive intervention by the State as a non-coercive regulatory element, through the use of a grant serving as an intergovernmental tax incentive. Such an incentive represents a strong extra budgetary economic instrument designed to achieve a constitutional preservation goal by promoting tax justice and influencing voluntary action by municipalities that desire an increase in revenues and an improved quality of life for their populations.

However, the Green ICMS's formula imposes on policy makers many challenges and presents some complexities considering the intrinsic consequences of any public policy that uses economic mechanisms to benefit the natural resources and biodiversity.

According to the Organisation for Economic Co-operation and Development (OECD, 1994), the redistributive effects of the use of economic instruments on environmental policies may constitute serious obstacles to their implementation, as potential 'losers' in the redistribution will be tempted to fight measures against their interests. Therefore, it is important to undertake a prior identification of possible redistributive effects, seeking to reduce them so as not to impede the implementation of the policy.

In relation to the Green ICMS, it is possible to observe that, by its very nature, the redistributive impacts are considerable, since the adoption of a new revenue distribution criterion implies a reduction in the value of the existing criteria. Incidentally, this is one of the purposes for adopting the system.[12]

Of course, this redistributive impact will make it difficult to reformulate the financial policy, as there will undoubtedly be those who will feel disadvantaged by the new pass-through criteria.[13] In this sense, attention should be paid to the fact that the larger, richer and more developed municipalities have a greater legislative representation and it is usually convenient for them to maintain the value added as the sole criterion for distribution of funds, which greatly hampers the possibilities for changes in the parameters for financial transfers. Therefore, an extremely important point is the careful selection of the new criteria to be adopted, as well as those that will be of reduced importance; decisions that definitively influence the extension of the redistributive effects of the system, facilitating, consequently, the political agreement as to its implementation.

In addition, it is very important to try to re-dimension the concept of development, separating it from the idea of environmental degradation, through a massive environmental education and information system, capable of generating new axiological concepts, in order to change fiscal parameters for a more harmonious coexistence with nature.

Another great challenge involves the transparency and knowledge of the policy by the municipalities. Several municipal managers are not aware of the institute's existence and how it works. Others who know complain about the lack of transparency in the methodology adopted to calculate the share of each municipality.

In the State of Pará, the current method used to calculate the quota shares is extremely complex. It is a factorial analysis that is a form of arrangement and verification of statistical data, unintelligible to operators of law and to society in general. This lack of transparency is a great obstacle to the full effectiveness of the Green ICMS.

Moreover, policy makers must consider that most of the territorial limitations that the municipalities undergo are imposed by State and/or federal determinations. This means that the municipalities have a very limited control of their territory such that, in effect, they end up being punished or rewarded by the conduct of the State or federal government. Municipalities do not have the power or the institutional framework to manage State or federal areas, and, in fact, 'few mayors are interested in municipality assuming responsibilities tied to natural resource management because doing so could affect important economic interests and, according to the local view, depress the municipality's economy' (Toni, 2003, p. 148).

Finally, the implementation of the Green ICMS needs to take into account the imbalance in tax revenue collection and how much they receive as intergovernmental financial transfers in Brazil's fiscal federalism. Data published by the Brazilian Treasury Department (Tesouro Nacional, 2017) revealed that, in 2016, Federal and State transfers corresponded to more than 75 percent of the budget in 82 percent of the municipalities in Brazil.

Even though Brazil's fiscal framework established by the 1988 Constitution already stipulated big intergovernmental transfers, this kind of instrument can always induce municipalities to underutilize their own tax bases (Shah, 1990) and become increasingly dependent on intergovernmental transfers.

These challenges reflect some of the difficulties involved in the implementation of the Ecological ICMS. They do not erase the importance of the policy as an innovative instrument to compensate for the opportunity cost of environmental protection, especially when we see the rise of Ecosystem Services Payments policies. Nevertheless, such policy deserves further and detailed analysis in order to guarantee its consistency and compatibility with several fiscal, environmental, constitutional and territorial complexities found within the federal universe in which it is applied.

## 6 CONCLUSION

Considering the importance of the various instruments in environmental policy, the Green ICMS appears to be an experience unique to Brazil in terms of environmental taxation (OECD, 2015).

The Ecological ICMS policy has now been in effect in some Brazilian States for a number of years, and it represents an economic extra budgetary mechanism that is aimed at achieving the constitutional purpose of environmental preservation, the promoting of tax justice, and influencing volunteer action of municipalities in search of sources for increased income. The policy specifically aims to motivate and compensate municipalities for investing in the environmental quality of its territories.

In this context, one thing that is noticeable is the beginning of a clear and simple way to align the economic systems with environmental preservation interests, providing incentives so that the municipalities will maintain environmental conservation areas without suffering excessively from losses that are accrued due to limited economic development.

This instrument has sought to correct distortions in the Brazilian federal fiscal system in order to unite the fight against federal inequality and environmental protection, facing, in order to achieve such objectives, many challenges. However, there is a need for a strong movement, both academic and governmental, focused on perfecting the institute. It is very important to have a progressive evolution driven by the increase in the amount of research on the subject in conjunction with partnerships between study groups and the government.

This policy, in spite of these challenges, presents great positive results that point to it having the capacity to be an instrument of effective tax justice and environmental protection in Brazil. Given its short implementation time, this economic instrument can therefore be considered to have been a success.

# NOTES

* This work benefited from the support of 'Investissements d'avenir' of the French National Agency for Research (Ceba, ref. ANR-10-LABX-25-01).

1. Brazil's Ministry of Environment has a high-resolution and interactive map where all the country's federal, State and municipal conservation units are displayed and can be accessed at: <https://mmagovbr-my.sharepoint.com/personal/22240033827_mma_gov_br/Documents/Site%20CNUC/A0_CNUC_PT-BR.pdf?slrid=5cd13a9e-4080-4000-ce27-f1a58288af00.>. The EMBRAPA (2013) and (2017) websites cited in the list of references also have a variety of maps that indicate conservation units.
2. The State of Paraná was the first to consider this reality. A study from Paraná's development bank suggested creating a fund to compensate municipalities with ecological protected areas (Ames and Keck, 1997).
3. It is important to add that, even though the effect of a Pigouvian intervention via state action has been challenged, according to François Lévêque (1996), by other theories such as economic theory of regulation by George Stigler, public choice theory by Gordon Tullock, and Ronald Coase, the concept of externalities still holds.
4. An example is the *Taxa de Controle e Fiscalização Ambiental* (environmental control tax), imposed by Instituto Brasileiro do Meio Ambiente e dos Recursos Naturais Renováveis (Federal Environmental Institute) over polluter activities.
5. This has already been the case in many States. In Paraná, for example, the number of municipalities benefiting grew from 122 in 1992 to 192 in 1998; in São Paulo, it increased from 104 (1994) to 152 (1999), configuring the ecological ICMS as the main source of revenue for a number of municipalities in all the States in which it is applied (Campos, 2000).
6. According to Loureiro (1998) the increase in these areas is one of the most important results of the Ecological ICMS project, since they represent an active participation of the private interest itself, mainly in the preservation of green areas in places whose forest cover practically no longer exists.
7. With the adoption of State Law N°. 12,040, of December 28, 1995 – known as the 'Robin Hood Law'.
8. The R$ is the symbol of the Brazilian currency called the Real.
9. This criterion was used, for instance, in the first State of Pará's legislation, according to article 4, item I, paragraph 'a' of State Decree No. 775 of 2013. This law can be accessed at: <https://www.semas.pa.gov.br/2013/06/27/d-e-c-r-e-t-o-no-775-de-26-de-junho-de-2013-publicado-no-doe-no-32-426-de-27062013/>.
10. State Law N°. 5.100 of 2007, article 2, paragraph 2 item II. This law can be accessed at: <http://alerjln1.alerj.rj.gov.br/CONTLEI.NSF/c8aa0900025feef6032564ec0060dfff/edd5f699377a00078325736b006d4012?OpenDocument>.

11. State Decree of Tocantins nº 1.666 of 2002, article 1º, item II, letter 'a'. This law can be accessed at: <http://dtri.sefaz.to.gov.br/legislacao/ntributaria/decretos/Decreto1.666.02.htm>.
12. Taking from the rich municipalities to give to poor municipalities is the motto adopted by the Law establishing the Ecological ICMS in the State of Minas Gerais, known as the 'Robin Hood Law'.
13. Through a study carried out in the States of Minas Gerais and Rondônia (Grieg-Gran, 2000), about 40 per cent of the municipalities experienced some loss with the adoption of Ecological ICMS. However, most municipalities (about 60 per cent), especially the poorest, experienced a significant increase in funding. This factor also meant that many municipalities with low conditions to increase primary production opted for the creation areas of environmental preservation as a means of increasing budget revenues.

## REFERENCES

Ames, B and Keck, M (1997), 'The Politics of Sustainable Development: Environmental Policy Making in Four Brazilian States' *Journal of Interamerican Studies and World Affairs*, 39(4), 1–40.

Campos, L (2000), 'ICMS Ecológico: Experiências nos Estados do Paraná, São Paulo, Minas Gerais e alternativas na Amazônia', http://www.mma.gov.br/port/sqa/ brasiljl/doc/icmspnp.pdf.

Domingues, J M (2012), 'Tax System and Environmental Taxes in Brazil: The Case of the Electric Vehicles in a Comparative Perspective with Japan' *Osaka University Law Review*, 59.

Drummond, J A, Franco, J L, and Ninis, A B (2009), 'Brazilian Federal Conservation Units: A Historical Overview of their Creation and of their Status' *Environment and History*, 15(4), 463–91.

Drummond, J A, Franco, J L, and Oliveira, D (2012), 'An Assessment of Brazilian Conservation Units – A Second Look' *Novos Cadernos NAEA*, 15(1) 53–83.

Embrapa (2013), 'Alcance Territorial da Legislação Ambiental, Indigenista e Agrária', https://www.embrapa.br/gite/projetos/alcance/index.html.

Embrapa (2017), 'Atribuição das Terras do Brasil', https://www.embrapa.br/gite/projetos/atribuicao/index.html.

Grieg-Gran, M (2000), *Fiscal Incentives for Biodiversity Conservation: The ICMS Ecológico in Brazil*, London: International Institute for Environment and Development.

IDESP (2013), *Contribuições ao debate na aplicação do ICMS Verde no Estado do Pará*, Belém: IDESP.

Lévêque, F (1996), *Externalities, Public Goods and the Requirement of a State's Intervention in Pollution Abatement*, Paris: CERNA – Centre d'économie industrielle Ecole Nationale Supérieure des Mines de Paris.

Loureiro, W (1998), *Incentivos fiscais para conservação da biodiversidade no Brasil*, Curitiba.

Loureiro, W (2008), *ICMS Ecológico, uma experiência brasileira de pagamentos por serviços ambientais*, Fundação SOS Mata Atlântica, Curitiba: The Nature Conservancy (TNC).

Marshall, A (1920), *Principles of Economics*, London: Macmillan and Co.

Nolte, C, Agrawal, A, Silvius K M, and Soares-Filho, V S (2013), 'Governance

Regime and Location Influence Avoided Deforestation Success of Protected Areas in the Brazilian Amazon' *Proceedings of the National Academy of Sciences of the United States of America*, 110(13), 2–6.

Nunes, T S S (2010), *A efetividade das unidades de conservação e das terras indígenas na contenção do desflorestamento na Amazônia Legal*, Belém: Federal University of Pará – Institute of Geoscience.

OECD (1994), 'Working Paper n° 92', http://www.oecd.org/dev/1919252.pdf.

OECD (2015), *OECD Environmental Performance Reviews: Brazil 2015*, Paris: OECD Publishing.

Oliveira, A C (2014), *ICMS Ecológico e Desenvolvimento: analise dos Estados de Rondônia, Tocantins, Ceará e Pará*, Belém: Cesupa.

Oliveira, A C and Tupiassu, L (2016), 'ICMS Verde para a redução do desmatamento Amazônico: estudo sobre uma experiência recente' *Veredas do Direito*, 13(25), 277–306.

Pigou, A (1932), *The Economics of Welfare*, London: Macmillan.

Riani, F (1997). 'O novo critério de repartição do ICMS aos municípios mineiros: avaliação dos resultados e sugestões' *Revista de Administração Municipal*, 221, 61–76.

Rosembuj, T (1995) *Los tributos y la proteción del medio ambiente*, Madrid: Marcial Pons.

Rubinstein, F (2010), 'Promoção da equalização fiscal no federalismo brasileiro: o papel dos fundos de participação' in: Conti, J M, Braga, C E F, and Scaff, F F (eds), *Federalismo Fiscal: questões contemporâneas*, Florianópolis: Conceito Editorial.

Scaff, F F and Tupiassu, L V C (2005), 'Tributação e Políticas Públicas: O ICMS Ecológico' in: Torres, H T (ed.), *Direito Tributário Ambiental*, São Paulo: Malheiros, pp. 724–48.

Seroa da Mota, R (2000), 'Forestry Taxes and Fiscal Compensation in Brazil', in: Rietbergen-Mccracken, J, and Abaza, H (eds), *Economic Instruments for Environmental Management: A Worldwide Compendium of Case Studies*, London: Earthscan/UNEP, pp 185–96.

Shah, A (1990), *The New Fiscal Federalism in Brazil.* World Bank.

Tesouro Nacional (2017), 'Balanço do Setor Público Nacional Exercício de 2016', https://www.tesouro.fazenda.gov.br/documents/10180/390400/BSPN+2016+-+Vers%C3%A3o+Final+-+sem+marca%C3%A7%C3%B5es.pdf/2db4fb40-516e-4d13-8cc8-ee6541e10aa8.

Toni, F (2003), *Municipal Forest Management in Latin America*, Indonesia: CIFOR.

Tupiassu, L V C (2006), *Tributação Ambiental: A utilização de instrumentos econômicos e fiscais na implementação do direito ao meio ambiente saudável*, Rio de Janeiro: Renovar.

Young, C and Roncisvalle, C (2002), *Expenditures, Investment and Financing for Sustainable Development in Brazil*, Santiago: United Nations.

# 11. Climate change-related action and non-productive investments in the European Union

**María Amparo Grau Ruiz**

## 1 INTRODUCTION

The purpose of this chapter is to analyse the innovative perspective adopted in the European Union (EU) to address environmental challenges through public support to finance climate action.[1]

The recent work carried out by the European Court of Auditors (ECA) contains useful findings related to several relevant issues. First, the 'mainstreaming approach', which has a widespread scope but complicates the effective control of environmental spending, will be described and assessed. Secondly, the concept of non-productive investments (NPIs) will be studied, since they promote sustainable use of agricultural land but it is necessary to adopt some rules to make them cost-effective.

At the moment, within the EU, the Sustainable Development Goal (SDG) 13 'climate action' trend cannot be calculated due to insufficient data over the past five years. On 20 June 2017 the Council adopted conclusions on 'A sustainable European future: The EU response to the 2030 Agenda for Sustainable Development' and called upon the Commission to carry out detailed regular monitoring of the SDGs at EU level.[2]

Two Special Reports presented by the ECA merit a thoughtful consideration[3] and the next pages will try to summarize their principal contents, highlighting the most attractive ideas. From a methodological perspective, this chapter will be focused exclusively on the dialogue between EU institutions. After reviewing the ECA's recommendations, and the European Commission's (EC) reactions, it is clear that enhancement of the 'quality' of public support is desirable in order to achieve a more responsible use of public funding in line with sustainable objectives.[4]

# 2 THE EU BUDGET ON CLIMATE ACTION

## 2.1 The Stated Budgetary Objective

There is a clear commitment at EU level: at least 20 per cent of EU budget for 2014–20 should be spent on climate-related action. However, there are some difficulties in checking what is happening in practice. The current strategy of 'mainstreaming' has led to the integration of actions tackling climate change into EU policies, instruments, programmes or funds (spending more on existing climate-related programmes, adapting them, or creating new ones and developing features). This has an advantage: an enormous coverage and sensitization of different sectors to care for climate action. Nevertheless, the main drawback appears in implementation, as the underlying system is quite complex.

Basically, the European Parliament and Council set targets in several main budget areas through the legislation governing the respective funding instruments. This can be done in several ways: in their preambles, by fixing binding targets, or setting mandatory minimum thresholds.[5]

The European Investment Bank (EIB) falls outside this scheme; its lending operations are outside the EU budget. Notwithstanding this, the EIB has set its own target: committing a minimum of 25 per cent of its total lending for climate action and a 35 per cent climate action share in its investments in developing countries.[6]

## 2.2 The Main Flows of the Planned Climate Funding

The European Structural and Investment Funds (ESIF) aim to reduce regional imbalances across the EU, with policy frameworks set for the seven-year Multiannual Financial Framework (MFF) budgetary period. ESIF comprise the European Regional Development Fund (ERDF), the European Social Fund (ESF), the Cohesion Fund (CF), the European Agricultural Fund for Rural Development (EAFRD) – aimed at helping the rural areas of the EU meet a wide range of economic, environmental and social challenges – and the European Maritime and Fisheries Fund (EMFF). In all of them, the funding for climate action has been integrated to a certain extent. In addition, mainstreaming has reached the EU research and innovation programme for 2014–20, Horizon 2020 and the LIFE Programme, that is the specific EU programme supporting environmental, nature conservation and climate action projects throughout the EU, currently for 2014–20.

For a better understanding of the nature of the problem, it is helpful to clarify the meaning of some expressions. The 'agricultural direct payments' are those granted to farmers in order to support their incomes and remunerate them for their production of public goods through 'greening', and in combination with 'cross-compliance'. Greening refers to agricultural practices beneficial for the climate and the environment (e.g. crop diversification); and cross-compliance designates a system which links most Common Agricultural Policy payments to farmers' compliance with basic rules for the environment, food safety, animal health and welfare, and good agricultural and environmental conditions of land.

The ESIF and agricultural direct payments account for around three-quarters of the planned climate funding, so initiatives affecting them have a huge impact. Both are areas under shared management of the EU and the Member States (MS), which implies greater complexity (as explained below).

### 2.3 The Content of the Audit and Its Main Results

The ECA makes an effort to look at whether the above-mentioned 20 per cent target has added value to the fight against climate change through a larger allocation of EU funds to climate action and/or better-focused EU funding for climate action. It examines whether there has been a quantitative or qualitative improvement.

The findings are not homogeneous. It is therefore necessary to distinguish among funds. Some advances are detected in the ERDF and the CF.[7] But this does not happen in the ESF,[8] in the areas of agriculture, rural development, fisheries,[9] and research.

### 2.4 The Methodological Arguments to Support the Criticism

Behind the inter-institutional conversation lie methodological concerns. The weightings given to EU funds going to projects, measures or actions are based on the Organisation for Economic Co-operation and Development (OECD) Rio markers. However, the EU uses climate coefficients of 0 per cent, 40 per cent and 100 per cent depending on the action's contribution to climate change mitigation or adaptation. The Commission deemed this OECD methodology inappropriate, particularly in the EAFRD field, because the measures have multiple and systematic co-benefits of an environmental, economic or social nature.

Nevertheless, the ECA stresses some weaknesses. On the one hand, in certain areas the conservativeness principle fails. This principle has been developed to avoid overestimates in climate funding. It points out that, if calculated in accordance with internationally established methodologies, the contribution from agriculture and rural development would be reduced by up to approximately 33 billion euros. On the other hand, the

ECA criticizes that there is no information on how much money is spent on climate change mitigation and adaptation; though as an exception climate tracking is only applied for external instruments.[10]

The Court denounces that in substantial parts of the EU budget, the absence of specific targets makes it difficult to meet the 20 per cent general target (e.g. in the area of competitiveness for growth and jobs). Moreover, the information on what funds plan to achieve or have achieved in terms of results (such as greenhouse gas (GHG) reductions) is only available for parts of the budget. Some data are not comparable, which hampers their consolidation.

In the case of the ERDF, another inconvenience is that MS are not obliged to use the common greenhouse gas reduction output indicator (except for energy investments). Furthermore, different methodologies and tools are used. Where indicators exist, they focus on mitigation within the energy sector.

In Horizon 2020[11] the Commission intends to apply compulsory climate-related checks to be performed on the whole work programme, prior to approval. In addition, climate-related considerations should be included in the standard project proposal templates and in the award criteria. Allocating a sufficient budget to measures that make a sizeable contribution to climate change objectives would help meet the target. The ECA admits that these are useful proposals, but claims that they do not present any quantifiable targets or models as to how the 35 per cent specific target might be met.

### 2.5 Changing Responsibilities According to Policy Area

When a policy area is in the direct management mode, the Commission itself manages the programme. This happens in most of Horizon 2020 and the LIFE Programme. Conversely, under the shared management mode, the Commission appraises and discusses objectives, actions and choice of indicators and approves the programming documents prepared and submitted by the MS. Consequently, the MS are responsible for designing, implementing and monitoring the measures included in the programmes. Responsibilities vary depending on the policy area.

From an institutional perspective at the EU level, each Commission Directorate General (DG) is responsible for incorporating climate action into its individual spending programmes, and for its implementation, in cooperation with the MS where appropriate. DG Climate Action provides knowledge and guidance on climate action to other Commission DGs and DG Budget coordinates the preparatory work for the allocation of resources to climate in the budgets, and collects and presents the related information. The Secretariat-General of the Commission plays a coordinating role, ensuring the overall consistency of the EU actions, which includes the mainstreaming policies.

## 2.6 Recommendations Made by the ECA and the EC's Reactions

### 2.6.1 Robust multiannual consolidation exercise

The Commission should carry out annually a robust, multiannual consolidation exercise to identify whether climate expenditure is on track to achieve the 20 per cent target. The current tracking method used by the Commission and MS indicates weaknesses in reporting and comprehensiveness. It does not reflect the full financial effects of EU spending on climate action, and it does not differentiate between mitigation and adaptation measures.

The Commission response is that it aimed to make the EU budget a pioneer in fostering mainstreaming, and that the MFF sets out a dedicated programme for addressing climate action (under the LIFE Programme).[12] It emphasises that the primary objective of different EU programmes is set in the respective legal base, and mainstreaming by nature means making the implementation of the primary objective consistent with EU climate policy. As the potential individual contribution of each fund varies according to its primary mission, the system to monitor it should differ. Of course, the tracking system of climate spending in the EU budget could be improved, increasing granularity, but the Commission does not foresee further disaggregation with regard to mitigation and adaptation due to its unclear administrative impact and to avoid double counting in case of co-benefits. The Commission disagrees with the ECA's choice to consider a set of principles agreed in 2015 by certain organizations that implement international development assistance as a benchmark for assessing the appropriateness of the mainstreaming approach. It adds that the co-legislators did not consider specific targets necessary for promoting climate mainstreaming in all programmes, although the Commission concurs that specific objectives could be useful for advancing climate mainstreaming. At this stage, the Commission does not consider the introduction of any additional reporting requirements to be necessary but recognizes that the link between spending, mobilized actions, and results should be improved. Some common indicators have been developed based on policy learning within specific programmes (e.g. an indicator measuring GHG emissions in the monitoring and evaluation framework of the Common Agricultural Policy). However, it is not technically possible to prepare consolidated data on reduction in GHG emissions from activities financed through the EU budget nowadays.

The Commission considers that a climate impact of 8 per cent[13] of non-greening direct payments is not overestimated. This is a proxy based on the penalty system of cross-compliance. Under shared management, the MS determine which priority or focus area a measure contributes to most. The variations in climate-relevant spending in different country programmes reflect the sector priorities of the country and should not be seen as an indication of broader or narrower commitment to climate action. With regard to the monitoring system to track the ESF contribution to climate action, the Commission adds that it has been strengthened in the 2014–20 financial period (by including climate action as a secondary theme).

### 2.6.2 Comprehensive reporting framework

The ECA asks the Commission to report, annually, consolidated information on the progress towards the overall 20 per cent target in its annual management and performance report and in each relevant annual activity report, including the details on progress on action plans where they exist. The information on the 'climate contribution of financial instruments' should also be reported. In the areas under shared management, the ECA asks the MS to report, in their annual implementation reports to the Commission, on the areas where there are potential opportunities for climate action, outlining how they plan to increase climate action. Both the Commission and the MS should ensure that data collection differentiates between mitigation and adaptation.

The Commission agrees to show relevant aspects of, and progress made on, climate action in the relevant Annual Activity Reports. It does not accept the recommendation to report on financial instruments, and makes a reference to the new annual management and performance report, which summarizes the EU budget performance.

### 2.6.3 Assessment of climate change needs

The Commission should ensure that plans are based on a realistic and robust assessment of climate change needs. This is important when planning the potential contribution to the overall target in climate action from individual budget lines or funding instruments.

The Commission agrees to consider climate change needs and the potential to contribute of different areas when proposing a new overall political target. However, it does not accept the recommendation to plan specific contributions for each area or programme.

### 2.6.4 Correction of overestimations

The ECA recommends correcting overestimations in the EAFRD by reviewing the EU climate coefficients set.

The Commission says that the tracking methodology needs to remain stable during the current MFF for reasons of predictability, consistency and transparency. However, the Commission will consider ways of fine-tuning the tracking methodology for the EAFRD for the post-2020 programming period, without increasing the administrative burden.

### 2.6.5 The drawing up of action plans

In the ECA's opinion, since there is a risk that the expected contributions from a particular area may not be achieved, the Commission should draw up an action plan setting out in detail how it expects to ensure the catch-up needed.

The Commission replies that it will assess opportunities to increase climate relevance in the context of the mid-term reviews of individual programmes and policies. Individual action plans would not be appropriate as individual programmes already provide processes for priority setting depending on the management mode. The Commission also believes that tracking commitment appropriations was a necessary first step and provides a reasonable proxy, balancing the administrative burden.[14] It admits the need to strengthen the focus on results of the EU spending. However, the current performance framework set by the legislator provides for climate-related indicators only for a part of the spending and they cannot be aggregated. In the set-up of the next MFF performance framework some changes could be made.

### 2.6.6 Development of indicators

In line with the 'budget for results' initiative, the development of indicators monitoring the actual spending on climate action and related results is also recommended to the Commission (in the area of shared management in cooperation with the MS). It should be a harmonized and proportionate system applicable in all areas that contribute to the achievement of the target, facilitating the exchange of good practices. In particular, the ECA advises the Commission to assess GHG emissions and reductions brought about through EU-funded measures.

The Commission does not accept the recommendation to develop a new monitoring system now because it would lead to an increase in the level of the administrative burden imposed on MS. It will strengthen and improve comparability of the climate-related results indicators in all EU budget areas and will consider options in the context of the next MFF to establish climate-related results indicators, in particular to assess GHG emissions and reductions. The Commission will continue to actively promote the exchange of good practices in this specific area (e.g. the Expert Group on Monitoring and Evaluating the Common Agricultural Policy). It supposes that the introduction of greening will contribute to additional climate actions. The green payment has de facto increased the baseline beyond which environmental practices can be paid under Rural Development Programmes (RDPs).[15]

### 2.6.7 Exploration of all potential opportunities

To ensure a real shift towards climate action, the ECA suggests that the Commission should identify areas with underutilized potential and develop action plans for increasing the climate action contribution (e.g. the ESF); and that the Commission and the MS should increase the mainstreaming of climate action by developing new, or retargeting existing measures in agriculture, rural development and fisheries.

According to the Commission's reply, changing the multiannual financial programming at this phase in ESIF programmes under shared management is neither practical nor feasible. Similarly within this programming period it is not possible to develop new climate action measures in the legislative framework in the EMFF.

## 3 THE SUPPORT FOR NON-PRODUCTIVE INVESTMENTS IN AGRICULTURE

### 3.1 Concept and Content of Non-productive Investments

An NPI is a financial incentive for the owners of agricultural holdings to undertake environmentally friendly investments. NPIs do not generate a significant return, income or revenue, or increase significantly the value of the beneficiary's holding, but they have a positive environmental impact. The EAFRD Regulation provided NPIs with a complementary role, but granted MS the discretion to choose the level at which this complementarity should occur.[16] MS should ensure it by explicitly linking NPI support to measures and/or objectives.

NPIs have varied content (e.g. restoration of landscape features, such as restoration of traditional dry-stone walls supporting terraced vineyards, or creating a habitat) in the framework of Axis 2 of the EU rural development policy with regard to the sustainable use of agricultural land. In the programming period 2014–20, the current measure 4.4 provides support for NPIs linked to the achievement of agri-environment-climate objectives. The beneficiaries are farmers and/or land managers. Their investments do not lead to any significant increase in the value or profitability of the agricultural holding.

### 3.2 The Main Concerns Expressed in the ECA's Audit[17]

The ECA has found that, although all MS defined their agri-environment needs in broad terms, they mitigated the risk of untargeted support by restricting the NPI types eligible for funding. It observes that there is indeed overall consistency between the type of NPIs selected by MS and the agri-environment needs described in their RDPs.

However, the new EAFRD Regulation contemplates 'integrated projects' concerning investments in physical assets, whereby the same beneficiary applies once for implementing actions under at least two different measures or sub-measures. This integration causes a risk of overlapping support if the activities to be funded under each measure are not clearly demarcated.

NPIs should address related environmental needs at a reasonable cost. As the proportion of the investment costs funded with public money is higher (up to 100 per cent) for NPIs than for other EAFRD investment measures, their beneficiaries may have less incentive to contain their costs. The issue of proportionality of the public aid rates applied for NPIs arises. Since almost any kind of investment can potentially provide a direct or indirect economic benefit, the applicable regulations state that NPIs should not provide a significant economic return to the beneficiary. However, the Commission did not provide guidance as to what a significant economic return is or how MS should assess it.

To safeguard the principle of economy, the remunerative characteristics should be considered by the MS when determining the amount of public NPI support. It is important to avoid overpayment of the related investment costs. These costs could also be lowered once the investment benefits are contemplated.

The MS should take into consideration the economic benefit, the agronomical benefit, the degree of financial assistance necessary to encourage undertaking the NPI and the basic support rate applicable to productive investments in the same geographical area when defining criteria.

In practice, some weaknesses in the selection procedures led MS to fund NPI projects, which were ineligible for EU funding. In Denmark the preliminary feasibility studies for wetland projects were treated as individual NPIs. Fewer than half of the financed projects were eventually followed by separate applications for actual investments in wetlands. The use of feasibility studies may entail the risk that other general costs, such as consultation fees or professional advice, may be labelled as feasibility studies.

In Italy the national authorities had defined that only farm holders registered in the farms register of the Chamber of Commerce could have access to NPI support. The national authorities did not verify compliance with this specific criterion independently. This task was entrusted to the producer associations, who had to certify that their affiliated members fulfilled the criterion. Some beneficiaries did not comply with this requirement and as such should not have accessed NPI support.

The project selection was insufficiently transparent in the United Kingdom. The national authorities used predefined lists of holdings prioritized according to their potential to deliver environmental outcomes followed by a negotiation procedure between the national authorities and the potential beneficiaries. However, information on the basis for the prioritization and assessment of the merits of the application was not available. In addition, the scheme documentation suggested that the inclusion of additional land management proposals was at the discretion of national officials.

Some weaknesses in the management and control systems of the MS led to reimbursement of unreasonably high or insufficiently justified investment costs (on the basis of unit costs higher than the actual market costs, or not appropriately verifying the reality of the costs claimed, or accepting the most expensive offer without comparing it against benchmarks). For example, until the end of 2013, the Danish authorities did not verify whether the invoices submitted had actually been paid prior to the payment request or whether deductible VAT, which is not eligible, was included in it. The lack of physical visits by the national authorities constitutes another shortcoming, especially considering the absence of a deterrent factor, such as significant financial participation by beneficiaries.

Performance information to show what has been achieved with the support to NPIs at EU and MS levels is lacking. There are no specific result indicators, and the MS did not define additional baseline indicators in relation to NPI support. It is essential to identify weaknesses in a timely manner so that the necessary corrective actions could be taken before the start of a new period. Analysing the causes of the irregularities detected through MS' own controls is also important. With the exception of Denmark, the national authorities made limited use of the irregularities found. They basically reduced the amounts to be paid to the affected beneficiaries. In addition to a regular monitoring, an evaluation should be made.

### 3.3 The ECA's Recommendations and the EC's Reactions

The Commission should monitor the relevant MS' implementation of NPIs through their annual implementation reports (with the number and proportion of NPI projects implemented in combination with other rural development measures or environmental schemes, including integrated projects). They should include in their evaluation plans an assessment of synergies in implementation.

The Commission agrees to encourage the relevant MS to report those data linked to the achievement of agri-environmental-climate commitments, where there is complementarity.

For the new programming period, the ECA says that the Commission should provide guidance to MS on selection criteria, having due regard to their transparency, and check they apply appropriate procedures. The MS should make public all criteria used in the selection and prioritization of NPIs and systematically verify the supporting documentation. They should also ensure an appropriate segregation of duties among organizations and persons involved.

The MS should also define appropriate criteria to determine the remunerative characteristics of NPIs benefiting from the highest aid rates and consequently modulate the intensity of support.

The Commission expresses in its reply that 'not providing a significant economic return' does not mean that no economic return can be accepted. The majority of investments include an element of return (e.g. planting hedges comprised of bushes which, when cut, can provide biomass). It is difficult to imagine a purely NPI. The Commission agrees that the 'productive' or remunerative features should be limited, recalling, on the downside, the administrative burden linked to extracting the economic benefits from the eligible expenditure. When the MS can make use of the simplified cost options for the calculations and payments of the grants, a fair and verifiable calculation must be carried out in advance.

Another problem may arise when the managing authorities reduce the aid intensity of the measure, even if environmental objectives are to be achieved. Sometimes, without a higher support rate, the agri-environmental-climate objectives cannot be met.

Regarding the risks of overlapping, the Commission explains that the objectives of measures can be set in RDPs, although in some cases setting a clear demarcation can be difficult and can lead to an increased administrative burden.

The Commission considers Denmark's intention to support feasibility studies under NPIs to be in line with the legislation. They may conclude that the investment will not contribute to the objectives of the support scheme. Even if they reach a positive conclusion on the investment, it may not be carried out for different reasons (e.g. budgetary constraints). An analysis on a case-by-case basis should be made.

The selection criteria are defined by the MS. Consultation follows with the Monitoring Committee (where the Commission is present in an advisory capacity). The Commission has produced guidance on eligibility and selection criteria to assist the MS and regions in the programming and implementation of the 2014–20 RDPs. The Commission in its audits verifies the use of selection criteria and has imposed financial corrections for deficiencies found.

Article 24(2)(d) of Commission Regulation (EU) No 65/2011 requires assessment from the MS on the reasonableness of costs during administrative checks using a suitable evaluation system, such as reference costs, a comparison of offers or an evaluation committee. This provision has been maintained with regard to the 2014–20 programming period. A balance needs to be found between what can be done through monitoring and evaluation, taking into account the risk of excessive administrative burden and the financial constraints. Collecting specific result indicators for NPIs can be burdensome. Moreover, the indicators are only the starting point for an evaluation. The evaluators can collect additional information (e.g. case studies).

The key reason behind the lack of sufficient information on performance in the mid-term evaluation is the timing (a minimum time is necessary to initiate the implementation; and it takes time for results obtained in operations, particularly in the environmental field, to become measurable).

The Commission carries out conformity audits in the MS to verify that the expenditure is in compliance with the rules. If weaknesses are found, financial corrections are applied. The audit work programme is determined on the basis of a risk analysis. The factors are financial importance, the quality of the control systems, the characteristics of the paying agency, the complexity of the measures and any other information from other bodies. Since the amount of funding for NPIs is relatively low, priority is given to more financially important measures.

The Commission has provided guidance, both for specific and horizontal measures, on eligibility conditions and selection criteria; and on investments, including NPIs. The managing authority and the paying agency must jointly *ex ante* assess the verifiability and controllability of all measures included in RDPs. Reporting has been based on a rationalized common monitoring and evaluation system (and the SWOT analysis of the RDP aims at establishing the baseline for this system).

## 4 CONCLUSION

There have been serious attempts to incorporate climate concerns into policy areas and the corresponding funds of the EU Budget. Efforts have been made for better-targeted spending, mainly by developing requirements for measures pursuing climate-related objectives.

The EU mainstreaming approach seeks to achieve greater climate relevance across operations financed by the EU budget. This pioneering EU method can be further improved by finding a balance between the robustness of data and the administrative effort required (both from the Commission and the MS).

The existing system of *ex ante* commitment tracking may be considered an efficient proxy because the expenditure data would not provide useful information (due to the time delay in moving from programming to expenditure). Any change in the composition of climate-relevant spending mirrors the changes in the spending priorities in individual programmes and the roles of different financing sources available at country level.

The European Funds and the national co-financing are to provide support to promote sustainable goals but always in a cost-effective manner. The EU MS are trying to adapt climate finance to current goals, as demonstrated by some good practices. More experience will probably be shared in the coming years. Meanwhile, the procedures are refined to provide information on real progress towards sustainability thanks to proper climate action funding. The lessons learned might be taken into account in the future design of tax expenditures.

## NOTES

1. European Environment Agency, Trends and projections in Europe 2016 – Tracking progress towards Europe's climate and energy targets, Publications Office of the EU, Luxembourg, 2016, https://www.eea.europa.eu/themes/climate/trends-and-projections-in-europe; European Environment Agency, Analysis of key trends and drivers in greenhouse gas emissions in the EU between 1990 and 2015, Publications Office of the EU, Luxembourg, 2017, https://www.eea.europa.eu/publications/analysis-of-key-trends-and/#parent-fieldname-title, accessed 5 February 2018.
2. 'For SDG 13 "climate action", data coverage is sufficient for the topic "climate mitigation", while trends of indicators on "climate impacts" and "climate initiatives" cannot be calculated due to insufficient availability of data. Indicators in the sub-theme "climate mitigation" predominantly show progress, with the EU being well on track to reach its targets for greenhouse gas emissions, renewable energies and energy consumption.' Eurostat, Sustainable development in the EU, Monitoring Report on Progress Towards the SDGS in an EU Context, 2017 edition, Publications Office of the EU Luxembourg, 2017, pp. 14 and 21.
3. ECA, Spending at least one euro in every five from the EU budget on climate action: ambitious work underway, but at serious risk of falling short, Special Report No. 31, Luxembourg, 2016. ECA, The cost-effectiveness of EU Rural Development support for non-productive investments in agriculture, Special Report, No. 20, Luxembourg, 2015.
4. The EU is committed to progress towards the SDGs, as shown in the Communication COM (2016) 739 Next steps for a sustainable European future, https://ec.europa.eu/europeaid/sites/devco/files/communication-next-steps-sustainable-europe-20161122_

en.pdf and the Accompanying Commission Staff Working Document on Key European action supporting the 2030 Agenda and the Sustainable Development Goals SWD(2016) 390 final, https://ec.europa.eu/europeaid/sites/devco/files/swd-key-european-actions-2030-agenda-sdgs-390-20161122_en.pdf, accessed 5 February 2018.

5. Regulation (EU) No 1291/2013 of the European Parliament and of the Council of 11 December 2013 establishing Horizon 2020 – the Framework Programme for Research and Innovation (2014–20) (OJ L 347, 20.12.2013, p.104). Regulation (EU) No 1305/2013 of the European Parliament and of the Council of 17 December 2013 on support for rural development by the European Agricultural Fund for Rural Development (EAFRD) (OJ L 347, 20.12.2013, p.487). Regulation (EU) No 1293/2013 of the European Parliament and of the Council of 11 December 2013 on the establishment of a Programme for the Environment and Climate Action (LIFE) (OJ L 347, 20.12.2013, p.185). Regulation (EU) No 1301/2013 of the European Parliament and of the Council of 17 December 2013 on the European Regional Development Fund and on specific provisions concerning the Investment for growth and jobs goal (OJ L 347, 20.12.2013, p.289).
6. 'Developed countries, including the EU and its MS, agreed in Copenhagen in 2010 to mobilize jointly USD 100 billion dollars a year in climate finance for climate action in developing countries. . . . The indicator measuring the contribution to the USD 100 billion international commitment on climate-related expending monitors the climate-related finance flows from EU MS, the EC and the EIB to developing countries'. Eurostat, Sustainable development in the EU, p.259.
7. The Commission estimated that 51.9 per cent of the climate funding planned under the 2014–20 MFF would come from agricultural funds.
8. Matching the supply of 'green workers' to the increased demand is crucial to the establishment of a green economy. The lack of qualified specialists in the green jobs sector will delay the implementation of climate-related actions. The ESF operational programmes explicitly allocated a very low share to climate action, specifically 1.4 per cent.
9. The estimated EMFF contribution to climate action in 2007–13 was 26 per cent of the Fund. There was a decrease in support for the replacement or modernization of engines and for the permanent or temporary cessation of fishing activities.
10. The 'Common Principles for Climate Mitigation Finance Tracking', developed by the joint climate finance group of Multilateral Development Banks and the International Development Finance Club, state: 'where data is unavailable, any uncertainty is to be overcome following the principle of conservativeness where under-reported rather than over-reported climate finance is preferable. The EIB, within its general principles for recording its climate action lending, refers to it in its "credibility" (in the case of doubt or uncertainty around climate impacts, the presumption will be to exclude). In addition, affirms that mitigation and adaptation are to be recorded separately.' Common Principles for Climate Mitigation Finance Tracking, 31 March 2015; European Investment Bank, EIB Climate Strategy, Mobilizing finance for the transition to a low-carbon and climate-resilient economy, 22 September 2015.
11. Council Decision 2013/743/EU of 3 December 2013 establishing the specific programme implementing Horizon 2020 (OJ L 347, 20.12.2013, p.965) states that the multiannual work programmes for Horizon 2020 must provide an indication of the amount of climate-related expenditure, where appropriate. Programmable actions have a clear thematic objective corresponding to one of the topics defined in the Commission's work programme. What share of bottom-up actions can be considered climate related depends upon researcher proposals, which are evaluated for their scientific excellence.
12. The data presented in the Staff Working Document (2016) accompanying the Midterm Review of the MFF (COM (2016) 603), show that over the 2014–20 period, a total of slightly above EUR 200 billion of climate-related spending will be reached. This is equivalent to around 18.9 per cent for the whole financial programming period.
13. The benefit of cross-compliance for climate actions can be assessed by applying a Rio

marker of 40 per cent to a share of 20 per cent of non-greening direct payments, as quantifying this contribution would be difficult and costly.

14. The current estimates are indeed done on the basis of programmed expenditure (draft budget); they are, however, updated based on the actual voted budget and amending letters. The Commission will assess the cost-effectiveness of providing *ex post* estimates based on actual commitments. Calculations based on the actual payments would create an additional administrative burden since the payments may last for years and may be subject to financial corrections.
15. 'Agri-environment-climate' refers to the new rural development measure 10 for the period 2014–20. The payments under this measure aim to preserve and promote necessary changes to agricultural practices with the intent to make a positive contribution to the environment and climate.
16. Recital 37 of the preamble to Council Regulation (EC) No 1698/2005 of 20 September 2005 on support for rural development by the EAFRD (OJ L 277, 21.10.2005, p. 1). Commission implementing Regulation (EU) No 808/2014 of 17 July 2014 laying down rules for the application of Regulation (EU) No 1305/2013 of the European Parliament and of the Council on support for rural development by the EAFRD (OJ L 227, 31.7.2014, p. 18). According to the EC SFC2007 system (May 2015) the total EAFRD and public (EAFRD plus national co-financing) expenditure for NPI (in euro) for the measure 216 regarding NPIs during 2007–13 are the following: Paid EAFRD 549.900.632 and public 859.331.286 euros; Programmed EAFRD 610.843.250 and public 1.013.996.313 euros. According to the EC SFC2007 system (March 2014) the implemented EAFRD funding for NPI in million euros was approximately 180 in the UK, 90 in Italy, and 20 in Portugal, Denmark and Sweden.
17. The audit focused on the 2007–13 programming period and covered the MS' management and control systems related to NPI support and a sample of 28 projects that reflected the most relevant NPIs in four MS: Portugal (mainland), Denmark, United Kingdom (England) and Italy (Puglia), which represented 80 per cent of EAFRD expenditure and 60 per cent of the beneficiaries of measure 216 at the beginning of the audit.

# 12. Total economic value of the Cagayan de Oro river basin

**Rosalina Palanca-Tan, Catherine Roween Chico-Almaden, Ma. Kresna Navarro, Marichu Melendez-Obedencio and Caroline Laarni Rubio-Sereñas**

## 1 BACKGROUND AND OBJECTIVES OF THE STUDY

The research aims to measure the total economic value (TEV) of the flow of ecosystem services from the Cagayan de Oro river basin (CDORB) in Mindanao, Philippines. The resulting estimates from the project provide the rationale for the adaptation of the river basin-wide payment for environmental services (PES) scheme which Xavier University-McKeough Marine Center (XU-MMC) is currently undertaking in collaboration with the Cagayan de Oro River Basin Management Council (CDORBMC). Through PES, financial resources can be generated and used to reward local initiatives that restore and preserve the ecosystem. This approach has been identified as one strategic way to safeguard and enhance the continuing flow of environmental services from the CDORB.

Estimates of the various TEV components of the CDORB ecosystem can provide the underlying basis and justification for the contributions that may potentially be collected from different economic sectors and social groups benefiting from the ecosystem's services. This TEV research looks at all potential groups of sources of rewards for the providers of services to rehabilitate and preserve the CDORB. As a well-protected watershed can provide security of water supply, fish supply, recreation, biodiversity, flood control and increased resilience to extreme weather events, the general public, especially those in the downstream communities, stand to benefit substantially. This project aims to go beyond the list of buyers already identified in the ongoing CDORBMC/XU-MMC PES program, namely: (1) Mindanao Authority (MinDA), a quasi-governmental entity mandated to promote, coordinate and facilitate the active and extensive participation of all sectors to effect socio-economic development in Mindanao, whose program Mindanao Nurture Our Waters (MindaNOW) supports the rehabilitation, protection and proper management of river basins and watersheds in Mindanao; (2) cooperatives such as MASS-SPEC Cooperative Development Center, the biggest regional federation of cooperatives in Mindanao with more than 300 primary cooperatives holding at least PhP11 billion assets with more than 1.5 million individual members; and (3) agribusinesses such as Del Monte and Agrinanas, whose operations are heavily dependent on the quality of the environment.

The research focuses on the following more numerous, long-term and sustainable buyers or sources of rewards for protectors of the CDORB:

1. The households in the downstream communities, in particular the city district and 40 barangays (17 urban and 23 rural barangays) in Cagayan de Oro (CDO) City, benefit from a well-functioning CDORB in terms of a stable supply of good-quality water, flood control, food supply (fish and other seafood), recreation (white water rafting and other water sports activities in CDO River and Macajalar Bay), power supply, climate change mitigation and biodiversity.
2. The industrial/commercial sector and other institutions account for a significant portion of water consumption in the area. Many of the industrial, commercial and institutional (schools, churches and government offices) water users have their own wells and are not served by water districts. Operations of these establishments are usually heavily dependent on a stable supply of water. They are also severely affected by extreme weather conditions and flooding. There has long been a clamor for these business and institutional establishments to contribute to watershed protection in the form of a raw water fee.
3. The fishing sector also stands to gain from a well-maintained watershed. Heavy siltation caused by denuded watersheds affects the productivity or catch of the fishermen around the CDO River and Macajalar Bay.
4. The tourism industry, which is now a major income-generating sector in the area, and still growing fast, is likewise heavily affected by siltation.

The identification of the beneficiaries of ecosystem services and the estimation of the values of the stream of benefits they derive from the CDORB will significantly facilitate the upscaling and acceleration of the implementation of the CDORBMC/XU-MMC PES program, which aims to reward the following upstream communities for their rehabilitation and preservation activities: (1) Miarayon Lapok Lirongan Tinaytayan Tribal Association (MILALITTRA), the resource managers working in the sub-watershed in Mount Kalatungan, and (2) the Kitanglad-wide Council of Elders with the Tribal Guards (Kitanglad Guard Volunteers) for Mount Kitanglad.

Economic valuation addresses the following environmental policy questions: (1) What is the value of the total flow of benefits from the CDORB? (2) How are these benefits distributed among the various stakeholders? and (3) Who can be potential financing sources for ecosystem conservation efforts?

It is also hoped that this research project, with all the surveys, key informant interviews, focus group discussions, research findings dissemination workshops and other activities it entails, can assist and contribute to CDORBMC/XU-MMC PES information campaigns, policy lobbying of PES/Ecosystem-based Adaptation (EbA) related laws in the local government units, and the development and integration of PES in the CDORB Master Plan.

## 2 THE STUDY SITE: CAGAYAN DE ORO RIVER BASIN

The CDORB has a total area of approximately 137,000 hectares spread over three provinces (Bukidnon, Misamis Oriental and Lanao del Norte), three municipalities (Baungon, Libona, and Talakag in Bukidnon) and two cities (CDO City in Misamis Oriental and Iligan City in Lanao del Norte).

CDORB lies between 124°0’39” and 125°21’39” latitude and 7°32’20” and 8°57’39” longitude. It is bounded by Northern Cotabato in the south, Lanao del Sur in the southwest, and Bohol Sea in the north. The highest elevations within the CDORB can be found at the peaks of Mount Kalatungan and Mount Kitanglad at 2,824 meters above sea level (masl) and 2,899 masl, respectively. The steep slopes are predominant in the upland area in the south and southeastern portion of the basin, where the majority of the river’s headwaters are located. They can also be found in the ridges of sub-basins where they serve as a topographic divide between subcatchments. Gentler slopes prevail along the coast and on the flat portions of several elevated terraces around the basin (CESM, 2014).

Cultivated areas mixed with brush land or grassland have the largest coverage, occupying 54 percent of the total basin area. Closed and open canopy of mature trees with a combined area of 346 km$^2$ accounts for

25 percent of the total basin area. The built-up areas, covering less than 1 percent of the basin, are found in the northern tip (CESM, 2014).

# 3 METHODS

A resource's TEV consists of use and non-use values. Use values from the CDORB ecosystem include benefits derived from a stable water supply, flood control, fishing and tourism. Non-use values, on the other hand, refer to resource values that are independent of people's present use of the resource. Non-use values may include existence values, bequest values and option values (Lockwood, 1998). Existence value is the value of simply knowing that the good exists, a sense of stewardship for preserving certain features of the natural resource. Existence values may accrue to both users and those who are not actually using the resource but who nevertheless have an interest in it. Bequest value is the value of being able to preserve the resource for future generations. Option value is the desire to preserve the resource for use in the future.

The particular methodologies employed to estimate the TEV components of the CDORB are as follows:

(1) For the use values (water supply, flood control, food supply, recreation) and non-use values (biodiversity) of the CDORB ecosystem to downstream households, mainly in CDO City, the study employs the contingent valuation method (CVM). This is a survey-based approach to valuing non-market goods such as public goods and environmental goods and services. CVM is now used extensively in both developed and developing countries to incorporate values of non-marketed services and amenities in public program assessment. Studies that employ CVM in estimating the total benefits from watershed rehabilitation and preservation programs include Alcon et al. (2013), Almanza and Martinez-Paz (2011), Perni et al. (2013) and Birol et al (2010). CDO households are asked to state their willingness to pay (WTP) or contribute to watershed rehabilitation and preservation efforts to ensure the steady flow of ecosystem services. Their WTP serves as an estimate of the benefits they perceive that they derive from the ecosystem's services. Both use and non-use values are integrated in a single estimate using CVM.

Systematic sampling procedure was employed in selecting the respondents in each barangay. The number of respondents in each barangay was set in proportion to the share of the barangay in the total CDO City population. Six bid levels were randomly assigned to respondents. The yes-no response to the dichotomous choice CVM question was analyzed using the framework developed by Hanemann (1984) based on the random utility model. A binary probit regression model was run to calculate parametric mean WTP. For the non-parametric estimate, the Turnball formula was employed. Individual households' WTP was aggregated to arrive at total WTP.

(2) The water supply for commercial, industrial and institutional establishments. This project utilizes results of groundwater depletion studies conducted in CDO in 2003 (Palanca-Tan and Bautista) and 2012 (Palanca-Tan). These previous studies produced estimates of the safe yield of the CDO aquifer, a survey of wells – Cagayan de Oro Water District (COWD) production wells and non-COWD deep wells, and estimates of the rate of groundwater withdrawal. The survey of deep well owners in the 2012 study included a WTP question for raw groundwater.

(3) For tourism income, value is estimated by collecting revenues and costs data from tourism business operators consisting of white water rafting companies for CDO River and resort/cottage owners for Macajalar Bay.

(4) Primary data collection for fishing income is likewise employed to arrive at net fishing value from the CDO River and Macajalar Bay. A sample of fishermen from different zones is surveyed for fishing effort, gear and catch to arrive at net annual fishing values.

The study area for CDO River is divided into subzones to determine the behavior of and trends in the fisheries of the different communities. These subzones represent upstream, midstream and downstream fishing communities. The upstream communities include Barangay Dansolihon of CDO City and two barangays of Talakag, namely Barangay 4 and San Isidro. The midstream communities include barangay Nicdao in Baungon and six barangays in CDO City (Kabula, Lumbia, Taguanao, Balulang, Balongis and Macasandig). The downstream communities include communities at the north end of the river, which borders the Macajalar Bay. These communities are all found in CDO City, particularly in Bonbon, Kauswagan, Consolacion, Burgos Street, Carmen and Upper Nazareth. Data in this study were collected through interviews of 455 fishermen. The fisheries of each subzone vary according to the number of fishermen and their resources as well as the topography of the river basin. The catch statistics specific to the river basin were also grouped according to the three subzones. To generate the total potential value of fish catch from the CDO River, revenues and costs of fish captured during peak and lean seasons are calculated and compared. The total fish value consists of the total fish caught, including those sold, given away and consumed by the fishermen's households. The costs involved in artisanal fisheries operations depend essentially on the fishing equipment. As for the labor cost, this can be estimated as the fishing effort, which is the number of fishing trips and the number of hours spent for each fishing trip during a given season. Maintenance costs of equipment are also minimal given the rudimentary type of equipment used. Thus, depreciation values are no longer included because often equipment is scrapped without salvage values. The estimated life span of equipment is one year as suggested by the fishermen.

To generate the total potential value from the fish catch in Macajalar Bay, calculation is based on the comparison of the potential values and costs of fish capture during the high and low months. The costs involved in fishing operations in the four barangays depend on the fishing equipment and vessels, labor and overhead costs. The study utilizes responses of 40 fishermen from the four barangays.

(5) The benefits from flood control are calculated using the damage costs function method, in addition to CVM. Existing estimates of damages from previous flooding, particularly from tropical storm Sendong (international name Washi), together with scientific projections of the frequency of these events with and without adequate watershed protection and management are cited. An estimate of the Value of Statistical Life (VSL) by Palanca-Tan (2014) is utilized to measure the life-saving benefit of flood control.

# 4 RESULTS

## 4.1 Household Benefits

Households, especially those in the downstream communities, benefit from the CDORB ecosystem in terms of a stable supply of good-quality water, flood control, food (fish) supply, recreational activities and biodiversity.

A total sample of 963 respondents was generated through face-to-face interview with the household head or the member making expenditure decisions in the family. Eighty-one barangays (40 barangays in the city central district and 41 other barangays), only 17 of which are rural, of CDO City were included in the sampling frame.

From the CVM survey results, the nonparametric estimate of the mean WTP for the CDORB rehabilitation and preservation program is 12.19 percent of the water bill. The parametric estimate using the coefficients of the regression result is 17.58 percent. With mean WTP per household ranging from 12.19 percent to 17.58 percent, and with a total household population of 137,465 (Philippine Statistics Authority, 2010) and a mean monthly water bill per household of PhP531.80, the total value of benefits from the rehabilitation and preservation of the CDORB would be PhP106,936,365.00–154,219,960.00 (US$2,371,094.00–3,419,511.00) per year.

### 4.2 Water Supply for Businesses, Industries and Other Institutional Entities

The water supply component of the use value of the CDORB accrues not only to households, but also to businesses and other establishments as well. An estimate of benefits derived by industrial, commercial and other institutions from the CDORB in terms of a sustainable supply of water presented in this subsection is drawn largely from studies undertaken by Palanca-Tan and Bautista (2003) and Palanca-Tan (2012) on groundwater extraction in CDO. The first, titled 'Metering and a Water Permits Scheme for Groundwater Use in Cagayan de Oro' (Palanca-Tan and Bautista, 2003) looked into the viability of metering groundwater extraction in CDO and collecting a raw groundwater fee to address two goals: (1) to control excessive abstraction of groundwater and (2) to generate revenues that can be used to preserve water catchment areas. The study found a strong WTP for raw water among the city's owners of groundwater supply systems, particularly businesses. It also found that payers want to see that revenues from the raw water fee are used to maintain and preserve the watersheds, to ensure a stable supply of water. The second study, 'Raw Water Pricing for Groundwater Preservation: A Policy Advocacy Exercise in CDO, Philippines' (Palanca-Tan, 2012), was an action research project undertaken in collaboration with the National Water Resources Board. The project endeavored to push the CDO government to legislate and implement a raw groundwater pricing scheme in the mode of PES in so far as the revenues from the raw groundwater fee would be earmarked for watershed rehabilitation and preservation programs.

Both studies presented estimates (the second study updating the estimates of the first study) of groundwater extraction by businesses and institutions and their WTP for raw groundwater. Overall, both the 2003 and 2012 surveys point to some WTP for groundwater resources among businesses and institutions in CDO to ensure a stable supply of good-quality water. In the 2003 survey, WTP ranged from PhP0.40–PhP10.00/ $m^3$ of groundwater withdrawn while in the 2012 survey the range was PhP1.00–PhP12.50/$m^3$. The 2012 study recommended a realistic value

PhP1.00/m$^3$ for the initial implementation of the raw groundwater pricing scheme. The same rate is charged by two water districts in the Philippines, namely the Laguna Water District and Metro Cebu Water District, on big commercial and industrial private (non-water district) groundwater abstractors in their service areas. The said rate is in the form of a production assessment fee (this rate was also the modal WTP, 18 out of 48 respondents, in the 2003 survey). Assuming a raw groundwater fee of PhP1.00/m$^3$ of water withdrawn and a monthly groundwater extraction of 978,209 m$^3$ by businesses and institutions in CDO, the total amount that can potentially be generated for watershed rehabilitation and preservation from this sector will be PhP11,738,508.00 (US$260,277.00).

## 4.3 Tourism (White Water Rafting) Value from CDO River

Early accounts of white water rafting activities in CDO River started in 2001. Full-blown commercial operation started in 2002. In 2015 there were seven companies in CDO City registered with the Department of Tourism (DOT) for the business of white water rafting. Currently, only six companies are active members of the Oro Association of Rafters (OAR). Data were obtained from the logbook keeper manning the reception area of the jump-off sites in Mambuaya, CDO City (for the beginner course) and Ugiaban, Talakag (for the advance course).

The recorded total number of tourists served for the beginner course in 2013 was 16,728 tourists. In 2014 the recorded total number of tourists served for the beginner course dropped by 27 percent to 12,250. This can be attributed to the first drowning incident, which killed one tourist on June 13, 2014. The incident resulted in massive cancellations of bookings as well as temporary termination of white water rafting activities two weeks after the incident. As for the first four months of 2015, it can be said that the industry had slightly recovered after the incident in 2014.

Data retrieved for the advanced course were dated January 2014 to February 2015. The difficulty in gathering information for this category can be attributed to the fact that the logbook is kept by the local people who do not have formal authorization from the government. They keep a record because they collect user fees of PhP10.00 per tourist and PhP40.00 per vehicle. When the whitewater rafting companies stopped helping the local people who man the jump-off point by collecting user fees, the recording also stopped. The recorded total number of tourists served in the advance course for 2014 is 7,637. The number of tourists for the advance course was estimated to be at least 60 percent of the number of tourists for the beginner course. There is a possibility that the figure might be lower for 2014 due to the same incident which killed a tourist. The recorded data for January and February 2015 are slightly lower relative to the data for the same period in 2014.

Based on the daily data gathered from August 2012 to April 2015, the estimated number of tourists served for the beginner course on a regular basis amounted to 16,397 annually. As for the advance course, it was estimated to be 9,682 annually.

Rafting companies record zero sales on an average of seven days in a month. This is equivalent to a total of 280 rafting days for each company per year. Average annual costs of operations per rafting company are about PhP591,634.13 (US$13,086.55).

In 2011 tariff rates for rafting were fixed by the Local Tourism Office of CDO City and were complied by all operators. The beginner course cost PhP700 per person, exclusive of meals, while advance courses (1 and 2) were PhP1,200 and PhP1,500 per person and extreme course PhP2,000.00. The revenue per person for the beginner course is PhP700 per trip while the average cost per person per trip is estimated at PhP526, resulting in a net revenue of PhP174. Given the average number of 16,397 tourists served annually, net revenue from the white water rafting beginner course is estimated at PhP2,853,078.00 (US$63,107.23). For the advance course, the net revenue per person is estimated at PhP624. With an average annual number of tourists of 7,637 the approximate total net revenue is PhP4,765,488.00 (US$105,407.83). Overall, the annual estimated total net revenue for both rafting courses is PhP7,618,566 (US$168,515.06).

There are indirect or induced benefits from white water rafting. Almost every group of rafting tourists would purchase photo and video packages. These services would entail an additional PhP500 per group. With an annual average rafting tourist groups for both beginner and advance courses of 4,873, photo and video income would amount to PhP2,436,500.00 (US$53,892.00).

Another indirect benefit is revenue from retail trade at jump-off sites. Retailers at jump-off sites are beneficiaries of the livelihood program of the Department of Labor and Employment Region 10. The project started in February 2014. The total revenue of the retailers from February 2014 to February 2015 amounted to PhP67,762.00. It was estimated that net revenue from total sales is around 30 percent.

The total estimated annual tourism value from both direct and induced activities from the whitewater rafting is PhP14.37 million (US$317,000.00).

### 4.4 Tourism Value (Resort/Cottage) from Macajalar Bay

One of the economic activities involving Macajalar Bay is resort/cottage-related trade. This section highlights the use value activities of resorts and cottage businesses in Barangay Bayabas and Bonbon and an estimate of these activities as a value of tourism in Macajalar Bay. The study involved interviewing three resort owners in Bayabas and three resort caretakers in Bonbon.

Revenue of the resort and cottage operation is obtained from room and cottage occupancy, entrance fees and rents for recreational facilities (e.g. videoke, volleyball and inflated ring). While room rates of resorts are generally constant throughout the year, their occupancy varies significantly between peak and off peak. This trend results in fluctuating revenues during the year.

The initial capital costs of a resort and cottage enterprise involve the construction materials for cottage and building and purchase costs of recreational facilities. Estimated monthly operating costs include maintenance cost of the building, inflated rings, videoke and volleyballs. To generate the total potential value of tourism activities around Macajalar Bay, the calculation is based on the comparison of the potential values and costs of resort and cottages during the high and low months. The procedure results in an annual net tourism value in Bayabas and Bonbon of PhP11,175,460.00 (US$190, 864.00).

### 4.5 Artisanal Capture Fisheries Value from CDO River

Fishing is unpopular as a livelihood in the CDORB. This may be due to the presence of better opportunities in other sectors in CDO City and the surrounding vicinities. The identified population of fishermen in all areas of the study site is estimated to be less than 10 percent of the total population. Artisanal fishing is an ancillary livelihood in most of the areas of study. The majority of the fishermen venture into other menial jobs due to proximity to the urban center. Many of them have subsidiary occupations which serve the dual purpose of alternative income and job opportunities and food source because fishing is seasonal. The peak fishing season usually spans two to four months. In rural communities of CDORB, most fishermen resort to farming as their subsidiary occupation at the onset of the rainy season during which fish catch tends to be lower.

The artisanal fisheries in all the communities are characterized by low capital investment and high labor-intensive practices. The fishing activities are usually carried out by traditional fishing methods such as gill nets, cast nets, long lines, hooks and line sets, arrows and rudimentary traps. Since the majority of the fishermen use rudimentary equipment, the resulting cost of acquiring and maintaining this equipment is relatively low. On average, a fisherman spends around Php3,000 a year on equipment.

The peak and lean seasons vary for all zones and even among communities within the same zone. The peak season in the downstream communities tends to coincide with the rainy season. According to fishermen in this area, this may be attributed to flooding, which increases the likelihood of fish from the upstream river sections being carried downstream. On the other hand, for most of the midstream and upstream communities, peak season coincides with the dry season because more fishing activities can be conducted when river water current is more manageable. The fishermen can dive into deeper channels of the river where potential fish catch is higher with slow water current. During the rainy season, full-time fishermen resort to the use of *besigan* or river traps to ensure continuous fish catch despite harsh river conditions.

The downstream area recorded an average weekly fish catch of only 5 to 10 kg of mainly smaller immature fishes. This fishing output suggests that this portion of the river has already been heavily fished. In the midstream communities, average fish catch is almost the same as the downstream, suggesting a similar situation. In the upstream communities, however the average fish catch is between 10 and 15 kg a week, suggesting relative abundance. In most instances, full-time fishermen's number of fishing trips per week ranges from four to six. This high fishing frequency is an indication of heavy dependence on fishing as a means of livelihood. During the lean months, fish catch tends to be lower by 40 percent for most of the fishing communities.

This study has identified 26 varieties of fish in all sampled communities. Each area tends to have a different concentration of fish variety. The concentration of fish in each zone tends to vary according to season. *Pigok* fish is the most dominant in most of the downstream and midstream communities, while *carpa* is the most abundant in the upstream communities. Certain varieties are abundant in particular zones while absent in others. For instance, *hipon* or *uwang* (shrimps) and *alimango* (crabs) are abundant in downstream and midstream zones but are scarce in upstream zones. This may be attributed to premature harvests in the downstream communities, which remove the fish before they can move upstream. Small shrimps and crabs in downstream areas are commonly used as bait, a practice which may be inefficient as shrimps, if matured and big enough, are more valuable (in the market) than the fish varieties for which the shrimps are used as bait.

Fish prices range from as low as PhP50 to as high as PhP400 at the fishermen's level. Usually, the fish vendor's price is higher by PhP20–100 depending on the fish variety. Because of the relative scarcity of most of the fish, prices do not vary much during peak and lean seasons. Fish prices also differ by zone: In rural areas most of the fish are cheaper by PhP20–100. According to the fishermen, the top-three most expensive fish are *damagan*, *pigok* and *balanak*. According to the Bureau of Fisheries and Aquatic Resources (BFAR), *pigok*, a fish variety found in big river systems, is the second-most expensive fish in the Philippines. Its price can go as high as PhP1,200 per kilogram. In CDO City, its price ranges between PhP500 and 800 per kilogram. Overall, the fishing sector in selected communities along the CDO River generates a total net value estimated at PhP24 million (US$543,000.00) annually.

## 4.6 Fishery Value from Macajalar Bay

The port of CDO serves as a major transfer junction for domestic and foreign trade to the province of Misamis Oriental, as well as an export outlet of the immense agricultural province of Bukidnon. Municipal waters within the jurisdiction of CDO City are divided into fishing zones. Fishing zone 1 comprises the fishing waters within the area from the mouth of the Iponan River, eastward up to the south of CDO River and extending towards the Macajalar Bay for three miles. Fishing zone 2 includes the fishing waters within the area from the mouth of the Bitan-ag Creek to the boundary of Bugo and Tagoloan and extending into the Macajalar Bay for three miles. The area between the mouth of CDO River and Bitan-ag Creek extending towards Macajalar Bay for three miles is zone 3, never opened to fish corrals, fishponds, culture beds or any construction or devices that will obstruct or hamper navigation. This study excludes fishing zone 2 and highlights fishery operation in four barangays (Bayabas, Bonbon, Lapasan and Macabalan). Among the selected barangays, Lapasan has the highest population, more than twice that of Macabalan (which ranks second among the four). Barangay Bonbon has the lowest population of 9,195 inhabitants. These barangays are observed to be heavily stirred by the brown coloration of the shore, which contributes to the low quality of the bay water.

The father and the adult male children are usually the full-time fishermen while young children and women are engaged in part-time fishing and fish trading. During high-catch months, household members whose primary occupation is fishing are engaged on municipal fishing for longer hours than during off-peak months. According to respondents, January to June are months with high fish catch while the next six months are either low fish catch or no fish catch at all. *Tamban* (Indian Sardine) is the fish species most frequently caught by fishermen. This species is listed as a prevalent fish species in Misamis Oriental (Philippine Statistics Authority, 2015). *Tamban* is caught in large volumes all year round but in relatively lower volumes during off-peak months.

Most fishing equipment of the fishermen is hand-made of nylon. Usually the smaller the holes of the fishing equipment and its firmness to hold fish catch, the greater its value to the fishermen. Depreciation values are no longer included because of the short life span of the equipment and the absence of salvage value. On the other hand, the license permit for the use of fishing equipment and vessel is also added in costs. As for the labor costs, these can be estimated from the wages in return for the effort expended by hired adults and children. It also includes the fishing effort of the fisherman measured in terms of the number of hours spent for each fishing trip in a season. Overhead costs consist of gasoline, ice, bait and transportation. The annual fishery value from Macajalar Bay is estimated at P62 million (US$1,378,741.00).

## 5 SUMMARY AND CONCLUSIONS

The study measures the TEV of the flow of ecosystem services from the CDORB Mindanao, Philippines. The study looks at all potential groups of buyers or sources of rewards/payments for the providers of services to rehabilitate and preserve the CDORB, namely households in the downstream communities, industries and institutional establishments, and the fishing and tourism sectors.

The total value of the benefits (stable supply of good-quality water, flood control, fishing and recreational value, biodiversity) that can be derived from the rehabilitation and preservation of the CDORB is estimated to be about US$5.0–6.1 million per year. The breakdown of this total into the different beneficiary sectors identified in the study is shown in Table 12.1.

These estimates of the various components of the TEV of the CDORB ecosystem can serve as the basis and justification for the contributions that may potentially be collected from different economic sectors and social groups benefiting from the ecosystem's services. Benefits derived by the different economic sectors from the CDORB warrant the watershed's preservation activities. A portion of these benefits can be collected to help finance conservation activities by the upland communities.

*Table 12.1 Economic values from the CDORB ecosystem services*

| Ecosystem services payer/ benefiting sector | Unit value | Total annual value |
|---|---|---|
| Use & non-use values (water supply, flood control, fishing, recreation, biodiversity) to CDO households | 12.19–17.58% of water bill PhP64.83–93.49 (US$1.44–2.07) per household per month | PhP106,936,365–154,219,960 (US$2,371,094–3,419,511) |
| Water supply for CDO businesses/institutions | PhP1.00 per m$^3$ of groundwater extracted | PhP11,738,508 (US$260,277) |
| CDO River tourism (white water rafting) income | PhP1,197,488 per month (US$26,552) | PhP14,369,857 (US$318,622) |
| CDO River fishing income | PhP4,482 (US$99.38) per fisherman per month | PhP24,470,665 (US$542,587) |
| Macajalar Bay tourism (resort/ cottage) income | PhP931,288 per month (US$15,905) | PhP11,175,460 (US$190,864) |
| Macajalar Bay fishing income | PhP8,481 (US$188) per fisherman per month | PhP62,181,227 (US$1,378,741) |

*Source:* Authors' calculations.

## REFERENCES

Alcon, F., J. Martin-Ortega, F. Pedrero, J.J. Alarcon and M.D. de Miguel (2013), 'Incorporating Non-market Benefits of Reclaimed Water into Cost-benefit Analysis: A Case Study of Irrigated Mandarin Crops in Southern Spain', *Water Resources Management*, 27, 1809–20.

Almanza, C. and J.M. Martinez-Paz (2011), 'Intergenerational Equity and Dual Discounting', *Environmental Development Economics*, 16, 685–707.

Birol, E., P. Koundouri and Y. Kountouris (2010), 'Assessing the Economic Viability of Alternative Water Resources in Water-scarce Regions: Combining Economic Valuation, Cost-benefit Analysis and Discounting', *Ecological Economics*, 69(4), 839–47.

Center for Environmental Studies and Management (CESM) (2014), 'Formulation of an Integrated River Basin Management and Development Master Plan for Cagayan de Oro River Basin', The Mindanao Development Authority, http://now.minda.gov.ph/?page_id=72.

Hanemann, M. (1984), 'Welfare Evaluations in Contingent Valuation Experiments with Discrete Responses', *American Journal of Agricultural Economics*, 66(3), 332–41.

Lockwood, M. (1998), 'Integrated Value Assessment Using Paired Comparisons', *Ecological Economics*, 25, 73–87.

Palanca-Tan, R. (2012), 'Raw Water Pricing for Groundwater Preservation: A Policy Advocacy Exercise in CDO, Philippines', in Kreiser, L., A.Y. Sterling, P.

Herrera, J.E. Milne and H. Ashiabor (eds), *Green Taxation and Environmental Sustainability*, Edward Elgar Publishing.
Palanca-Tan, R. (2014), 'Value of Life Estimates for Children in Metro Manila', in Barrett, S., K.-G. Maler and E.S. Maskin (eds), *Environmental and Development Economics: Essays in Honour of Sir Partha Dasgupta*, Oxford University Press, pp. 334–52.
Palanca-Tan, R. and G. Bautista (2003), 'Metering and a Water Permits Scheme for Groundwater Use in Cagayan de Oro', EEPSEA Research Report.
Perni, A., J.M. Martinez-Paz and F. Martinez-Carrasco (2013), 'Assessment of the Programme of Measures for Coastal Lagoon Environmental Restoration Using Cost-benefit Analysis', *European Plan Studies*, 21(2), 131–48.
Philippine Statistics Authority (2010), *2010 Philippine Statistical Yearbook*, National Statistical Coordination Board.
Philippine Statistics Authority (2015), *2015 Philippine Statistical Yearbook*, National Statistical Coordination Board.

# PART IV

# Incentives for the electricity sector

# 13. Low-income households in New York's Reforming the Energy Vision

**Ross Astoria**

## 1 INTRODUCTION

New York's Reforming the Energy Vision (REV) is an extensive overhaul of the regulatory framework governing the state's electrical power system. New York's regulatory commission for the electrical power sector, the Department of Public Service (NYDPS, or the Commission), began REV in 2014. The REV program now involves no fewer than 11 ongoing and interlocking dockets, each consisting of staff reports, Commission orders, intervenor comments, and subsequent tariff filings from each of the state's investor-owned utilities (IOUs).[1] It also includes rate cases, pilot projects, and coordination with the state's wholesale electrical market operator, the New York Independent System Operator (NYISO). In broad terms, NYDPS aims to align the utility's regulatory framework with the technological attributes of the distributed energy resources (DER) and thereby hasten the transition to a low-carbon electrical power system.[2]

The regulatory framework of the past hundred years, here labeled the *Hope* regulatory framework after the Supreme Court case from which it originates, supported the development of large, centralized fossil-fired generation which served a passive and expanding customer base. The physical hydrocarbon infrastructure developed under the *Hope* regulatory framework subsumed the poor into the utility's mass customer base so as to stabilize the utility's territorial monopoly and to preserve the utility as a site of capital accumulation. With a few notable exceptions, such as the Navajo Nation, access to electrical power in the United States is nearly ubiquitous, especially when evaluated against what one might expect given the racial and socio-economic striation characteristic of the United States.[3] The technological aspects of many renewable energy technologies contradict both the legal and physical infrastructure of the hydrocarbon electrical grid.[4] The carbon-free, micro-grid is emerging as a real technological possibility. As the old hydrocarbon grid comes under pressure from the energy transition, government will face a choice of how to incorporate the poor into both the legal and physical aspects of the renewable energy grid. Energy apartheid and energy democracy represent extreme points of a continuum. This chapter examines the REV effort to this point.

In Sections 2 to 4, I describe the structure of energy poverty governance under the old hydrocarbon infrastructure supported by the *Hope* regulatory framework. In the remaining sections, I describe a few of the salient challenges governments face in transitioning the poor to renewable energy networks and how the NYDPS has negotiated those challenges to this point. REV's efforts offer lessons to other jurisdictions.

## 2 CAPITAL, THE POOR, AND THE *HOPE* REGULATORY FRAMEWORK

States established railroad commissions during the 1880s and assigned them the responsibility of setting reasonable rates. The railroad industry soon challenged the constitutionality of those commissions and the rates they set and secured a ruling from the Supreme Court that the Takings Clause of the Fifth Amendment was applicable to the railroad's capital, which was legally identified as being invested for a public purpose. As such, any rates set by railroad commissions insufficient to allow the capitalist to receive just compensation would represent an unconstitutional taking of property. The culminating case is *Smyth v. Ames*, in which the Court found that

> a state enactment, or regulation made under the authority of a state enactment, establishing rates for the transportation of persons or property by railroad that will not admit of the carrier earning such compensation as . . . is just to it and the public, would deprive such carrier of its property without due process of law.[5]

The Court articulated a 'fair value' doctrine, which provided indeterminate guidance on how courts should evaluate the reasonableness of rates:

> in order to ascertain that value, the original cost of construction, the amount expended in permanent improvements, the amount and market value of its bonds and stock, the present as compared with the original cost of construction, the probable earning capacity of the property under particular rates prescribed by statute, and the sum required to meet operating expenses, are all matters for consideration, and are to be given such weight as may be just and right in each case.[6]

The early years of this constitutional protection were confused by incommensurate accountings of a utility's 'fair value,' and various groups attempted to secure valuations of the franchise monopolies favorable to their interest. Both the constitutional protection and the vague 'fair value' doctrine were made applicable to the electrical power industry, which became the site of a 40-year tussle of legal politics.[7] The Supreme Court's 1944 ruling in *Hope Natural Gas* settled the indeterminacy of the 'fair value' doctrine by allowing the utility's prudently invested capital duly depreciated to be accounted as its fair value for purposes of rate-making.[8]

Under the *Hope* regulatory framework, when a public service commission approves rates and tariffs, it first calculates the utility's 'rate base.' Per the prudent investment standard, the rate base is the value of the utility's fixed capital, duly depreciated. To this is added a reasonable return on investment for capital, generating the utility's 'revenue requirement.' The rates to be paid by the utility's customers are then computed so as to meet that revenue requirement.[9] To expand their ever-depreciating rate base, then, utilities must continually increase, or at least refresh, their capital expenditures, and this gives rise to the well-known capital expenditure bias, or capex bias.

The *Hope* rate-making process prioritizes capital's return on investment and thereby configures the franchise monopolies as sites of differential capital accumulation.[10] Among other effects, the *Hope* regulatory framework squeezes third-party capital, which does not enjoy constitutional protection, off the grid. The capex bias also prioritizes centralized, capital-intensive generation and infrastructure to the detriment of DER. Economic discourse frequently treats the capex bias as a sort of a natural fact, which it is not. Rather, the capex bias is an artifact of the *Hope* regulatory framework and overcoming it is one of the central aims of the REV.[11]

Under the *Hope* regulatory framework, a utility secures from the state a monopoly franchise to serve a particular geographic territory. Concurrent with the struggle over the appropriate valuation under the 'fair value' doctrine, the duration, extent, and service requirements of these monopoly franchises were all matters of political contestation during the formative years of the electrical power industry.[12] Salient to the governance of the poor, politicians modified or required the franchise monopolies to extend service to all within their franchise territory.[13] Under legal challenge from the utilities, the courts found that the franchise monopolies included a duty to serve all within their territory.[14] Being in the utility's territory, the duty to serve subsumed the poor into the *Hope* regulatory framework.

The utility's cost of fulfilling the obligation to serve the poor does not fall on capital, but upon ratepayers through utility-run affordability programs. In REV's low-to-moderate-income docket, for instance, the

NYDPS considered a proposal to require IOUs to institute arrears forgiveness programs.[15] In partially rejecting the proposal, the NYDPS noted that both bad debt and forgiven arrears are accounted in the utility's rate base as uncollectible revenue, so that those costs are shuffled onto other ratepayers. Since no data then available indicated whether such shuffling of costs was efficient, the NYDPS allowed arrears forgiveness programs 'for utilities who see value, but not presently require them for all companies.'[16] Under New York REV's reconfiguration of the state's utility low- and moderate-income docket, the total budget for utility-level affordability programs across all New York utilities for both gas and electric is $247.7 million (details below).[17]

## 3 LIHEAP: TAXPAYER-FUNDED LOW-INCOME PROGRAMS

The second policy which subsumes the poor into the energy governance structure of the *Hope* regulatory framework is the tax-financed Low-Income Home Energy Assistance Program (LIHEAP). The Community Service Administration (CSA) administered LIHEAP's predecessor, the Emergency Energy Conservation Program of 1975, which was geared toward weatherization.[18] In 1977, Congress added a 'Special Crisis Intervention Fund' which the CSA could use to make direct payments to fuel providers on behalf of the poor on an emergency basis.[19] These emergency funds were directed toward those states where utilities had been or were threatening to shut off service.[20]

The allocation of these 'Special Crisis' funds to states depended upon three variables: the coldness of the state's climate, households in poverty (with a special variable for elderly in poverty), and the relative cost of fuel.[21] In 1978, Congress further expanded the criteria to include other energy emergencies, such as a natural disaster, fuel shortages, and widespread unemployment.[22] The 1984 LIHEAP reauthorization modified these governance variables by including cooling degree days (CDD) and heating degree days (HDD) as variables and separating out low-income households from households generally.[23]

To receive a LIHEAP block grant from the Department of Health and Human Services, states must have established the bureaucracy for administering the program. In New York, the Office of Temporary and Disability Assistances (OTDA) administers the New York Home Energy Assistance Program (HEAP).[24] Under OTDA regulations, LIHEAP funds are further disbursed to county-level social service offices for their direct disbursal and administration.

Total LIHEAP funding jumped to around $4.5 billion during the Great Recession and has since settled back into annual appropriations of around $3 billion per year.[25] For the three years 2009 to 2011, New York's grant averaged about $484 million per year. For the subsequent years through 2017, grants to New York hovered between $350 and $380 million per year. Of the 50 states, New York receives the largest annual grant and Pennsylvania receives the second largest, of about $203 million in 2016.[26] Between LIHEAP and utility-funded programs a total of about $610 million is spent annually in New York on low-income assistance.

## 4 ENERGY POVERTY GOVERNANCE: ENERGY APARTHEID OR ENERGY DEMOCRACY

The *Hope* regulatory framework, then, configures the utility as a site of capital accumulation. The duty to serve subsumes the poor into this accumulation regime through a 'cross-subsidy' from the mass affluent to the poor. The poor thereby acquire access to sufficient amounts of heat and electrical power to contain the threat of political destabilization posed by pauperism while preserving the exchange-value of the utility's securities against the threat of excessive non-payment. Both utility affordability programs and LIHEAP, then, are charity programs which maintain political stability without disturbing status quo social and economic power asymmetries. The *Hope* accumulation regime, for instance, supposes that utilities will own generative assets whose productivity will be passively consumed by ratepayers. Neither charity program challenges this asymmetric and hierarchical allocation of generative assets.

The *Hope* accumulation regime, however, played a role in extending systems of electrical power and heat across the United States. Federal New Deal projects and programs, such as the Tennessee Valley Authority (TVA) and rural electrical cooperatives, did most of the heavy lifting in extending services to rural areas.[27] Further, the Federal House Authority's finance requirements for home mortgages extended electrical wiring into the 'average' home, and a series of New Deal federal loan guarantee programs enabled homeowners to finance appliance purchases, both of which underpinned utilities' post-war load growth.[28] Nonetheless, without the duty to serve, along with the cross-subsidies between economic classes which support that duty, the extension of the grid within territories served by the IOU franchise monopolies would likely have fractured along existing geographies of economic and racial segregation.

The transition to non-fossil electrical power systems, then, seems to offer up two general governance courses departing from the *Hope* regulatory framework. One is energy apartheid. Along this course, the price of renewable technologies becomes economically attractive, either on their own or with the support of Pigouvian carbon pricing. Nonetheless, the upfront or 'overnight' costs will remain a financial barrier for many individuals and communities, so that the development of green micro-grids would proceed along the geographic striation of wealth in America (and globally). 'Gated' communities would erect renewable micro-grids and 'defect' from the fossil macro-grid. With fewer of the mass affluent available to support utility low- and moderate-income (LMI) affordability programs, capital would be deprived of the internal cross-economic class subsidization which characterizes those programs. Utility capital would become increasingly exposed to deteriorating security exchange-values while the poor would become increasingly exposed to an unreliable fossil macro-grid. Under energy apartheid, the poor are ejected from the *Hope* regulatory framework and abandoned.

The second, preferred course is to govern the transition into a system of energy democracy. Under energy democracy, individuals and communities would themselves own generative energy assets. The surplus energy generated by those assets would recirculate back into the community and support the community's capital accumulation, rather than that of the utilities' shareholders. To prevent the poor from being ejected and abandoned by the *Hope* regulatory framework, states would refashion the regulatory framework to support the inclusion of the poor in the emerging physical and governance structures of renewable energy networks.

Energy democracy contradicts the *Hope* regulatory regime in at least two ways. First, since a utility's revenue requirements are based upon its depreciated capital expenditures, the decentralized ownership of generative assets undermines the utility's accumulation regime when an entity other than the utility owns those assets. Energy democracy cuts into a utility's rate base, erodes its revenues, and thereby disfigures the accumulation regime. Second, since the objective of energy democracy is the ownership of generative assets by those formerly beneficiaries of utility affordability programs, the utility would no longer be serving the poor, who fall out from under the utility's governance regime. To the degree that service to the poor provides an ideological plinth to the notion that utility capital is invested for a public purpose, their removal from that regime undermines the reasoning which places utility capital under constitutional protection. From the point of view of the transition to renewables, the governance question is whether states can overcome these contradictions and facilitate the emergence of energy democracy and avoid energy apartheid.

## 5 NEW YORK'S STRATEGY: HARMONIZING EXISTING PROGRAMS

The NYDPS's early strategy on LMI households is two-fold. First, the Commission harmonized existing programs in a docket addressed specifically to LMI households.[29] Second, in a series of sub-rulings throughout the other dockets, the NYDPS is attempting to develop governance structures and finance mechanisms which make DER accessible to LMI households while configuring DER as a site of capital accumulation for 'third-party' financial capital.

In the LMI docket of 2016, the Commission ordered the state's utilities to develop discount rates for LMI customers based on the percentage of a household's income devoted to purchasing energy, referred to as a household's 'energy burden.' The Commission identified 6 percent as the target energy burden, which is 20 percent of the maximum 30 percent shelter cost used as a benchmark for mortgage financing.[30]

The Commission ordered utility LMI affordability programs to be tier-based, fixed discounts which lattice off of the state's HEAP program. New York's HEAP program adds on benefits to households which are below 130 percent of federal poverty level (extreme poverty) or which contain an individual who is over 60, under six, or permanently disabled (vulnerable individual). Both the extreme poverty add-on and the vulnerable individual add-on increase the HEAP benefit by $25 per month. Utility LMI assistance programs are to identify a tier for each add-on (as a proxy for need), and calculate their assistant programs accordingly.[31]

The OTDA uses 60 percent of the state's standard median income (SMI) to determine HEAP eligibility. New York's SMI is $58,003, 60 percent of which is $34,802 or $2,900 a month. A 6 percent monthly energy burden, then, is $174. To this is added the HEAP benefit of $25, tallying to $201 maximum energy burden under the lowest tier, and so forth for the higher tiers.[32]

Under the *Hope* regulatory framework, the cost of utility affordability programs is borne by other ratepayers, and the Commission deemed it appropriate to establish a funding limit of 2 percent of electric and gas revenues.[33] While 'the cost of the program should be borne by all classes of customers,' the Commission left that allocation for individual rate cases.[34]

The 6 percent energy burden will require the enrollment of 2.3 million New York households into the utility LMI programs, up from about 1.1 million under the previous programs.[35] Of the New York utilities, Con Edison's $69.5 million budget is the largest for administering the new LMI program, an increase of about $10.5 million from its old programs. Orange and Rockland's new costs show the greatest increase, from about $4.5

million to $20.2 million. The total budget for utility assistance programs across the New York utilities is $247.7 million for both gas and electric, an increase of 87 percent from the previous budget of $132.4 million.[36]

## 6 NEW YORK'S STRATEGY: DER GOVERNANCE AND FINANCE REGIME

The NYDPS's second strategy is to move LMI households to a new governance regime, with two main components. The first is to shift LMI customers into distributed energy resource networks and the concurrent legal institutions which support those networks. The second is to use public finance through the NY Green Bank to support distributed energy networks which both serve LMI households and operate as sites of third-party capital accumulation.

Transitioning to a renewable energy grid poses a variety of challenges, two of which are made particularly salient in the context of LMI households. The first is misalignment between infrastructure and the built environment on the one hand, and the technological attributes of renewable energy resources on the other. Zoning ordinances, building codes, and mortgage financing requirements of the post-WWII era all confederated to produce a built environment reliant upon the consumption of high-energy fossil fuel.[37] Much of this reliance was not accidental. The natural gas distribution industry, for instance, supported building codes which required the use of natural gas as a strategy to secure customers.[38] Likewise, electrical utility campaigns attempted to increase loads through the promotion of the 'modern' electric home and appliances, and infrastructure planning locked in the combustion engine as the near-exclusive mode of transportation.[39] Hence, while the *Hope* regulatory framework configured the franchise monopolies as a site of capital accumulation, infrastructure construction and building codes developed the consumers' concurrent dependency upon the utilities' generative and transmission assets. Renewable technologies will require an infrastructure and built environment similarly oriented around the technological attributes of those technologies. As a simple example, the efficient installation of solar photovoltaic (PV) requires either a south-facing roof or a use right in real estate. Because of this misalignment between the built environment and the technological attributes of renewable technologies, building codes and infrastructure planning are central features of any greenhouse gas mitigation plan.[40] While the configuration of the built environment poses a general barrier to the transition, it is more acute for LMI households, many of which are renters.[41]

New legal institutions are also required to allow households to own renewable generating assets which are distant from the loads they serve. The REV Community Distributed Generation order (CDG) and the Community Choice Aggregation (CCA) order contain some of the main alterations to achieve this end.[42] The CDG docket establishes the constitutive rules under which communities may build energy projects. It sets out what percentage of the project may be owned by a single entity, how exported energy will be credited to those who own a subscription in the project, and the various responsibilities of the project sponsor with respect to both subscribers and the utility.[43] The solar garden is the paradigmatic but not only such project supported by the CDG order. The CDG rules are generally applicable, so project sponsors do not need to seek regulatory approval of each project. CDG is a radical departure from the *Hope* regulatory framework, which preserved the utilities' monopsonies. As a matter of contrast, it is very difficult to build solar gardens in some states because the utilities block the state commission from developing the tariffs and interconnection rules necessary for solar gardens to access the grid.

The CCA docket allows municipalities to create a separate legal entity authorized to contract for energy services for the entire municipality. Individual members of the municipality would then be able to opt out of the service. In this way, an existing legal entity – the municipality – is 'repurposed' to overcome the organizing challenges which otherwise face communities interested in concerted energy development projects. The NYDPS has explicitly noted that it plans for the CDG and CCA dockets to take up some of the governance of the LMI households, insofar as LMI households could own a share in a community energy project.[44] Hence, the emerging REV regulatory framework seems to be shifting LMI households into the physical structures and legal institutions of renewable energy networks.

The second barrier to the development of DER is financing. This general challenge is exacerbated by the *Hope* regime's capex bias, itself made even more acute in the context of LMI households which have no 'discretionary' income, much less capital. Hence, in the CDG order of 2015, the Commission requested staff to prepare a study on the financing option available to LMI households. The initial conclusions were not promising. For instance, each of the potential financing options available in the 'green energy space' posed substantial barriers to the participation of low-income households. Owning the asset, leasing the asset, and contracting a power purchase agreement with financing from a third party all required either a suitable credit score or a suitable debt-to-income ratio. Moreover, DER is a 'new asset class' about which financiers still needed training before figuring out how to extract from it a return on investment. Reliance upon the NY Green Bank and community development financial institutions (CDFI) seemed to offer the most promise.[45]

When the NYDPS released its Value of Distributed Energy Resources (VDER) order in March of 2017, it expressed frustration over the lack of LMI participation in community distributed generation (CDG) projects.[46] The Commission therefore convened a second working group, which met over the summer and fall of 2017 to confer on LMI participation in community energy projects.

Staff published the results of this working group in mid-December 2017.[47] After identifying the challenges LMI households face in participating in DER, including renting rather than owning and lack of access to capital, the staff advanced three main recommendations. The first is to develop a 'Bill Discount Pledge (BDP) Program' which would rely 'on the funding stream associated with utility low-income affordability bill discount programs. The BDP Program would allow low-income customers to redirect all or a portion of their discount to investments in CDG projects, thereby reducing or eliminating the need for subscription payments by low-income subscriber.'[48] The second is to direct public capital through the NY Green Bank, a division of the New York State Energy Research and Development Authority (NYSERDA), to support LMI projects. Created by the NYDPS in a December 2013 order, the NY Green Bank is capitalized through system benefit charges.[49] The Bank uses a chapter approach to organize its investments in the clean energy transition and its most recent Low-to-Moderate Income Chapter includes a new Low-Income Community Solar initiative. 'Under the Low-Income Community Solar initiative, NYSERDA will develop a CDG subscription model specifically for low-income customers, with subscriptions offered either at no cost, or where low-income participants pay a portion of their saving to the sponsor to receive other benefits.'[50] The third staff proposal is to develop a loss reserve program in which 'public funds would be held in reserve to cover potential losses that project owners and lenders may incur, if low-income CDG subscribers default on or terminate CDG subscriptions at a higher rate than other customers.'[51]

Following the recommendation of the Pace Energy and Climate Center, the staff report also discussed an 'Environmental Justice Location Incentive.'[52] In the VDER, the Commission identified the components for valuing the contributions of DER to the grid.[53] Because those components are graphically represented as stacked upon each other in a column, it is referred to as the value stack. The Pace proposal would add an additional component to the value stack when the project was located in an Environmental Justice Community. The Value Stack Working Group is considering the proposal, which would, at the minimum, require a method of identifying Environmental Justice Communities and a cost recovery mechanism.

As of mid-Winter 2018, the NYDPS had not acted upon any of the proposals in the staff report.

## 7 CONCLUSION

The transition to renewable technologies will decentralize the electrical grid's physical and governance structures. Cross-economic class subsides subsumed the poor into the *Hope* regulatory framework in support of a regime of capital accumulation while managing the threat pauperism poses both to political stability and to the exchange-value of the utility's securities. The decentralization of the fossil grid might eject LMI households from the *Hope* regulatory regime, setting governments on paths toward either energy apartheid or energy democracy. Which way does REV face?

The December 2017 staff report on LMI orients the policy program. First, the BDP Program would convert the utility affordability payments and taxpayer LIHEAP funding from charity programs to 'forward' subsidies. These forward subsidies would enable LMI households to own generative assets, a component of energy democracy. Second, in a manner similar to how New Deal loan guarantees made appliances affordable to newly wired households, the NY Green Bank's Low-Income Community Solar initiative shows promise in combining public money with third-party capital to overcome the regulatory and organizing barriers which hinder access to community DG. Third, the loss reserve program would protect capital's return on investment, perhaps overcoming the *Hope* capex bias and mobilizing third-party private capital into the green energy space. This would combine energy democracy with a regime of 'green' capital accumulation. Finally, the Environmental Justice Location Incentive might support an equitable transition against the inertia of a country striated by race and wealth. Guarding against optimism, the REV regulatory framework seems to nucleate programs which conspire toward energy democracy.

## NOTES

1. NYDPS, Order Establishing a Community Distributed Generation Program and Making Other Findings, 15-E-0082 (2015); NYDPS, Order Establishing the Benefit Cost Analysis Framework, 14-M-0101 (2016); NYDPS, Order Authorizing the Clean Energy Fund Framework, 14-M-0094 (2016); NYDPS, Order Adopting a Clean Energy Standard, 15-E-0302, 16-E-0270 (2016); NYDPS, Order Adopting a Ratemaking

and Utility Revenue Model Policy Framework, 14-M-0101 (2016); NYDPS, Order Adopting Distributed System Implementation Plan Guidance, 14-M-0101 (2016); NYDPS, Order Adopting Low Income Program Modifications and Directing Utility Filings, 14-M-0565 (2016); NYDPS, Order Authorizing Framework for Community Choice Aggregation Opt-Out Program, 14-M-0224 (2016); NYDPS, Order Resetting Retail Energy Markets and Establishing Further Process, 15-M-0127, 12-M-0476, 98-M-1343 (2016); NYDPS, Order Authorizing Utility-Administered Energy Efficiency Portfolio Budgets and Targets for 2016–2018, 15-M-0252 (2016); NYDPS, Order on Net Energy Metering Transition, Phase One of Value of Distributed Energy Resources, and Related Matters, 15-E-0751, 15-E-0082 (2017).

2. Ross Astoria, 'On the Radicality of New York's Reforming the Energy Vision.' *The Electricity Journal*, no. 30 (2017): 54–8.
3. Andrew Needham, 'Power Lines: Phoenix and the Making of the Modern Southwest.' In *Politics and Society in Twentieth-Century America*, edited by Gary Gerstle, William Chafe, Linda Gordon, and Julian Zelizer. Princeton and Oxford: Princeton University Press, 2014; Thomas Piketty, Emmanual Saez, and Gabriel Zucman, 'Distributional National Accounts: Methods and Estimates for the United States.' National Bureau of Economic Research, 2016.
4. Ross Astoria, 'Incumbency and the Legal Configuration of Hydrocarbon Infrastructure.' In *The Political Economy of Clean Energy Transitions*, edited by Channing Arndt Douglas Arent, Mackay Miller, Finn Tarp, Owen Zinaman. Oxford: Oxford University Press, 2017; Astoria, above n. 2.
5. *Smyth v. Ames*, 169 U.S. 466 (1898).
6. Id. at 526.
7. Barbara Fried, *The Progressive Assault on Laissez Faire: Robert Hale and the First Law and Economics Movement.* Cambridge, Mass.: Harvard University Press, 2001.
8. *Federal Power Commission v. Hope Natural Gas Co.*, 320 U.S. 591 (1944).
9. Lowell E. Alt Jr., *Energy Utility Rate Setting*. Milton Keynes: Lightning Source UK Ltd, 2006.
10. Tim Di Muzio, *Carbon Capitalism: Energy, Social Reproduction, and World Order*. London: Rowman Littlefield International, 2015.
11. Astoria, above n. 2.
12. Mark Granovetter and Patrick McGuire, 'The Making of an Industry: Electricity in the United States.' In *The Laws of the Markets*, edited by Michael Callon, 147–73. Oxford: Blackwell Publisher/The Sociological Review, 1998; Patrick McGuire, 'Instrumental Class Power and the Origin of Class-Based State Regulation in the U.S. Electrical Utility Industry.' *Critical Sociology* 16, no. 2–3 (Summer–Fall 1989): 181–204; Joseph Patrick Sullivan, *From Municipal Ownership to Regulation: Municipal Utility Reform in New York City, 1880–1907*. New Brunswick: Rutgers, The State University of New Jersey, 1995; Delos F. Wilcox, *The Indeterminate Permit in Relation to Home Rule and Public Ownership*. Bulletin No. 35, Chicago: Public Ownership League of America, 1926.
13. Scott Hempling, *Regulating Public Utility Performance: The Law of Market Structure, Pricing and Jurisdiction.* Chicago: American Bar Association, 2013.
14. *New York ex re. N. Y. & Queens Gas Co. v. McCall*, 245 U.S. 345 (1917).
15. NYDPS, Order Adopting Low Income Program Modifications and Directing Utility Filings (2016).
16. Id. at 34.
17. Id. at Appendix C.
18. Libby Perl, 'The LIHEAP Formula: Legislative History and Current Law.' Congressional Research Service, 2012.
19. Id. at 3.
20. Id.
21. Id.
22. Id. at 4.

23. Id.
24. Social Service Law, §97.
25. U.S. Department of Health and Human Services, *LIHEAP Clearinghouse* (2018), available at https://liheapch.acf.hhs.gov/Tribes/graph_funds.htm.
26. Id.
27. Ronald R. Kline, *Consumers in the Country: Technology and Social Change in Rural America*. Baltimore and London: The Johns Hopkins University Press, 2000; Philip Selznick, *TVA and the Grass Roots: A Study in Politics and Organization*. Berkeley: University of California Press, 1984.
28. Ronald C. Tobey, *Technology as Freedom: The New Deal and the Electrical Modernization of the American Home*. Berkeley: University of California Press, 1996.
29. NYDPS, Order Adopting Low Income Program Modifications and Directing Utility Filings (2016).
30. Id. at 8.
31. Id. at 17–24.
32. Id. at 23.
33. Id. at 28.
34. Id. at 29.
35. NYDPS, Order Adopting Low Income Program Modifications and Directing Utility Filings, 3 (2016).
36. Id. at Appendix C.
37. David Nye, *Consuming Power: A Social History of American Energies*. Cambridge, MA: MIT Press, 1999.
38. Id. Mark Rose, *Cities of Light and Heat: Domesticating Gas and Electricity in Urban America*. University Park: The Pennsylvania University State Press, 1995.
39. Matthew Huber, *Lifeblood: Oil, Freedom, and the Forces of Capital*. Minneapolis: University of Minnesota Press, 2013; Harold Platt, *The Electric City: Energy and the Growth of the Chicago Area, 1880–1930*. Chicago: The University of Chicago Press, 1991.
40. California Air Resource Board, 'California's 2017 Climate Change Scoping Plan,' chapter 4, p. 73 (2017).
41. NYDPS Staff, Report on Low-Income Community Distributed Generation Proposal, p. 4.
42. NYDPS, Order Establishing a Community Distributed Generation Program and Making Other Findings (2015); NYDPS, Order on Net Energy Metering Transition, Phase One of Value of Distributed Energy Resources, and Related Matters (2017).
43. NYDPS, Order Establishing a Community Distributed Generation Program and Making Other Findings, pp. 7–17 (2015).
44. Id. at pp. 22–27.
45. NYDPS Staff, Proceeding on Motion of the Commission to Examine Programs to Address Energy Affordability for Low Income Customers, 14-M-0565 (2017).
46. NYDPS, Order on Net Energy Metering Transition, Phase One of Value of Distributed Energy Resources, and Related Matters (2017).
47. Staff Report on Low Income Customers.
48. Id. at p. 21.
49. NYDPS, Order Establishing New York Green Bank and Providing Initial Capitalization, 13-M-0412 (2013).
50. Staff Report on Low Income Customer, 28.
51. Id. at 31.
52. Id. at 32 *et seq*.
53. NYDPS, Order on Net Energy Metering Transition, Phase One of Value of Distributed Energy Resources, and Related Matters (2017).

# BIBLIOGRAPHY

Astoria, Ross. 'Incumbency and the Legal Configuration of Hydrocarbon Infrastructure.' In *The Political Economy of Clean Energy Transitions*, edited by Channing Arndt Douglas Arent, Mackay Miller, Finn Tarp, and Owen Zinaman. Oxford: Oxford University Press, 2017.

Astoria, Ross. 'On the Radicality of New York's Reforming the Energy Vision.' *The Electricity Journal*, no. 30 (2017): 54–8.

California Air Resource Board. *California's 2017 Climate Change Scoping Plan.* 2017.

Fried, Barbara. *The Progressive Assault on Laissez Faire: Robert Hale and the First Law and Economics Movement.* Cambridge, Mass.: Harvard University Press, 2001.

Granovetter, Mark and Patrick McGuire. 'The Making of an Industry: Electricity in the United States.' In *The Laws of the Markets*, edited by Michael Callon, 147–73. Oxford: Blackwell Publisher/The Sociological Review, 1998.

Hempling, Scott. *Regulating Public Utility Performance: The Law of Market Structure, Pricing and Jurisdiction.* Chicago: American Bar Association, 2013.

Huber, Matthew. *Lifeblood: Oil, Freedom, and the Forces of Capital.* Minneapolis: University of Minnesota Press, 2013.

Kline, Ronald R. *Consumers in the Country: Technology and Social Change in Rural America.* Baltimore and London: The Johns Hopkins University Press, 2000.

Lowell E. Alt., Jr., *Energy Utility Rate Setting.* Milton Keynes: Lightning Source UK Ltd, 2006.

McGuire, Patrick. 'Instrumental Class Power and the Origin of Class-Based State Regulation in the U.S. Electrical Utility Industry.' *Critical Sociology*, 16, no. 2–3 (Summer–Fall 1989): 181–204.

Muzio, Tim Di. *Carbon Capitalism: Energy, Social Reproduction, and World Order*. London: Rowman Littlefield International, 2015.

Needham, Andrew. 'Power Lines: Phoenix and the Making of the Modern Southwest.' In *Politics and Society in Twentieth-Century America*, edited by Gary Gerstle, William Chafe, Linda Gordon, and Julian Zelizer. Princeton and Oxford: Princeton University Press, 2014.

NYDPS. 'Order Establishing New York Green Bank and Providing Initial Capitalization.' 2013.

NYDPS. 'Order Establishing a Community Distributed Generation Program and Making Other Findings.' 2015.

NYDPS. 'Order Adopting a Clean Energy Standard.' 2016.

NYDPS. 'Order Adopting a Ratemaking and Utility Revenue Model Policy Framework.' 2016.

NYDPS. 'Order Adopting Distributed System Implementation Plan Guidance.' 2016.

NYDPS. 'Order Adopting Low Income Program Modifications and Directing Utility Filings.' 2016.

NYDPS. 'Order Authorizing Framework for Community Choice Aggregation Opt-out Program.' 2016.

NYDPS. 'Order Authorizing the Clean Energy Fund Framework.' 2016.

NYDPS. 'Order Authorizing Utility-Administered Energy Efficiency Portfolio Budgets and Targets for 2016–2018.' 2016.

NYDPS. 'Order Establishing the Benefit Cost Analysis Framework.' 2016.
NYDPS. 'Order Resetting Retail Energy Markets and Establishing Further Process.' 2016.
NYDPS. 'Order on Net Energy Metering Transition, Phase One of Value of Distributed Energy Resources, and Related Matters.' 2017.
Nye, David. *Consuming Power: A Social History of American Energies*. 1999.
Perl, Libby. 'The LIHEAP Formula: Legislative History and Current Law.' Congressional Research Service, 2012.
Piketty, Thomas, Emmanual Saez, and Gabriel Zucman. 'Distributional National Accounts: Methods and Estimates for the United States.' National Bureau of Economic Research, 2016.
Platt, Harold. *The Electric City: Energy and the Growth of the Chicago Area, 1880–1930*. Chicago: The University of Chicago Press, 1991.
Rose, Mark. *Cities of Light and Heat: Domesticating Gas and Electricity in Urban America*. University Park: The Pennsyvlania University State Press, 1995.
Selznick, Philip. *TVA and the Grass Roots: A Study in Politics and Organization*. Berkeley: University of California Press, 1984.
U.S. Department of Health and Human Services. 'LIHEAP Clearinghouse.' 2018. https://liheapch.acf.hhs.gov/Tribes/graph_funds.htm.
Staff, NYDPS. 'Staff Report on Low-Income Community Distributed Generation Proposal.' 2017.
Sullivan, Joseph Patrick. *From Municipal Ownership to Regulation: Municipal Utility Reform in New York City, 1880–1907*. New Brunswick: Rutgers, The State University of New Jersey, 1995.
Tobey, Ronald C. *Technology as Freedom: The New Deal and the Electrical Modernization of the American Home*. Berkeley: University of California Press, 1996.
Wilcox, Delos F. *The Indeterminate Permit in Relation to Home Rule and Public Ownership*. Bulletin No. 35, Chicago: Public Ownership League of America, 1926.

# 14. Mitigating the environmental consequences of electricity sector 'lock in': options for a decarbonised energy future

**Rowena Cantley-Smith**

## 1 INTRODUCTION

Electricity generation is one of the quickest-growing forms of final *end use* energy supply worldwide. With global electricity generation increasing almost fourfold during the last four decades the stationary energy sector has expanded to meet the ever-increasing demand for access to electricity. The reasons for this are hardly surprising given electricity's application across a diverse range of energy services, e.g., heat, light, power, transport.[1] In response to the growing global expansion of coal-fired generation construction projects, fossil fuels have continued to dominate the electricity generation fuel mix. With much of the increased electricity supply being produced through coal-fired generation, the adverse environmental consequences of secondary energy end use have also continued to rise. As the International Energy Agency (IEA) recently stated: '$CO_2$ emissions from energy account for the largest share of global anthropogenic GHG emissions, representing over three quarters of emissions from Annex I countries, and about 58 percent of global emissions.'[2] That is not, however, the totality of the problem. Over and above existing $CO_2$ emissions from fuel combustion, carbon-intensive infrastructure continues to characterise developments in this sector. In some instances, this involves decisions to extend the life of existing heavy polluting infrastructure, or engage in developments that fail to take account of new technologies and/or alternative energy sources, e.g., building inefficient, coal-fired power plants, and overly heavy dependence on natural gas. In this way, $CO_2$ emissions are being *locked in* for the expected lifetime of the stationary energy sector's infrastructure, often well beyond recognised emissions reduction target periods. The consequences of inadequate global response to this looming problem are considerable. Not only does stationary energy sector *lock in* threaten the likelihood of achieving adequate greenhouse gas emissions (GGEs) mitigation in coming years, but more so, delaying action now raises the prospect of substantially higher costs to do so in the future. Accordingly, this chapter examines the issue of $CO_2$ emissions lock-in ($CO_2$ lock-in) in electricity generation. By way of background, the enduring role of fossil fuels in global energy markets and associated levels of energy sector GGEs are précised at the outset (Section 2). Key features of stationary energy-sector $CO_2$ lock-in are then examined (Section 3). Existing policy and legislative responses of the European Union (the EU) and Australia are used to illustrate the divergent approaches to overcoming this potential barrier to a future decarbonised energy sector – dealing with existing (locked-in) emissions and seeking to avert the 'lock-in of emissions from future capacity'[3] (Section 4). Concluding comments complete this chapter's discussions (Section 5).

## 2 ENERGY AND CLIMATE CHANGE: THE GLOBAL PREVALENCE OF FOSSIL FUELS

### 2.1 Rising Global Energy Supply and Demand

In the past four decades, global demand for energy sources has escalated. According to the IEA's recent *Key World Energy Statistics*, global total primary energy supply (World TPES) and final energy consumption (World FEC) – expressed in terms of million tonnes of oil equivalent (Mtoe) – have more than doubled in the past 45 years. Between 1973 and 2015, World TPES has increased from 6,101 to 13,647 Mtoe, and World FEC from 4,661 to 9,384 Mtoe.[4] Over the same period, electricity generation (World E-GEN) – expressed in terms of terawatt hours (TWh) – has more than quadrupled from 6,131 to 24,255 TWh.[5] In addition to rising supply, generation, and consumption, there has been a notable shift in the relative contributions (expressed in percentage terms) of individual fossil fuels to the various energy fuel mixes. Between 1973 and 2015, fossil fuels' overall shares in energy fuel mixes have fallen, most notably in World E-GEN, where oil's share of the stationary energy fuel mix has fallen from 24.8 to 4.1 per cent.[6] Even so, the combined contribution of fossil fuels has barely changed over more than four decades, with oil's reduced fuel mix share offset by increased uptake of natural gas and coal.[7] Regardless, these relatively small changes are dwarfed by the overall growth in fossil fuels' supply and use between 1973 and 2015: (i) World TPES increased from 5,290 to 11,109 Mtoe; (ii) World FEC rose from 3,381 to 6,287

Mtoe; and (iii) World E-GEN rose almost fourfold, from 4,611 to 16,081 TWh.[8] Regional and national shares of World TPES, World FEC, and World E-GEN, have also shifted from OECD to non-OECD countries.[9] Most notably, between 1973 and 2015, the OECD's shares of World TPES and World FEC fell, while China's shares increased from 7.0 to 21.9, and 7.8 to 20.4 per cent respectively.[10] Similarly, China's 24.3 per cent share of regional electricity generation in 2015 is an enormous increase on its 1973 share of 2.9 per cent.[11] The majority of China's increased electricity generation to date has relied on coal, but renewable energy is assuming an increasingly important role in the national energy fuel mix.[12]

### 2.2 The Fossil Fuel Legacy: A Global Snapshot

Despite worldwide efforts directed towards stabilising global temperatures and protecting the global climate system from harmful human activities, more than four decades of energy data – précised above – show that, in respect of fuel combustion $CO_2$ emissions, efforts to date are failing to reign in global GHG emissions. The IEA's most recent data show that global GGEs from fuel combustion have more than doubled since 1973, from 15,458 to 32,294 Mtoe $CO_2$ in 2015.[13] Fossil fuels generate almost all of these $CO_2$ emissions.[14] Given that coal is 'nearly twice as emission intensive' as oil and gas,[15] it is not surprising that global $CO_2$ emissions have increased to such a large degree in the past four decades as a result of increased reliance on this highly polluting fossil fuel. Moreover, the shift in global energy demand mentioned previously is reflected in a similar shift in the source regions of $CO_2$ emissions.[16] During the last four decades, the Organisation for Economic Co-operation and Development's (OECD) contribution to global $CO_2$ emissions has almost halved.[17] Comparatively, China and Asia's respective shares of global GHG emissions have escalated dramatically, with China producing just over 28 per cent of the energy sector's $CO_2$ emissions in 2015.[18]

## 3 ELECTRICITY GENERATION: FOSSIL FUELS AND GGEs

It is hardly surprising that the global combined energy sector constitutes the single largest contributor to GHG emissions. The international community's response to this global problem has been discussed across a wide range of academic, governmental, industry and other literature. This need not be repeated here, save to say that this state of affairs has led to significant changes in international law, with the creation of international climate change law treaties obliging States to, *inter alia*, establish and maintain national emissions data, subjecting States to increasingly higher emission mitigation targets, and requiring implementation of mitigation and adaptation responses at national levels as the primary way to meet these targets.[19] Concluded and adopted in 2015, the Paris Agreement – and its forward-thinking approach to widening both the responsibilities and opportunities for addressing GGEs through State and non-State action – entered into force on 4 November 2016.[20]

### 3.1 Growing Demand for Electricity and $CO_2$ Lock-in

The electricity generation sector is in the midst of 'one of the most profound transformations since its birth in the late 19th century'.[21] In the ten years between 2001 and 2011, for example, 'new coal-fired generation covered nearly half of increased demand for electricity'.[22] More recently, the IEA reiterated coal's ongoing dominance in electricity generation, remarking that coal-fired generation still provides 'the backbone of the global power system (around 40 percent of global electricity supply)'.[23] Specifically, between 2006 and 2016, China and India accounted for most of the global expansion of the coal-fired generation construction projects – 86 per cent.[24] At one level, electricity consumers' rising demand is completely understandable, for instance:

> [Electricity] offers a variety of services from mechanical power to light, often in a more practical, convenient, effective and cleaner way than alternative forms of energy. For some applications, such as electronic appliances, electricity is the only option. In addition, electricity produces no waste or emissions at the point of use and is available to consumers immediately on demand (where service is reliable) without any need for storage.[25]

On another, the high dependence on fossil fuels – especially coal in electricity generation as discussed previously – means that as demand for electricity grows, so too do the environmental, social, economic and technical challenges associated with this form of stationary energy supply. Indeed, although 'the need for reliable and affordable power supplies has never been greater', as the IEA observes, the power sector's status as 'the single largest source of greenhouse-gas emission' has resulted in it being the 'principal focus of efforts to tackle climate change'.[26] Accordingly, addressing fossil fuels' continued dominance in electricity generation and the associated problem of $CO_2$ lock-in is essential to address rising GGEs and climate change. In general terms, commentators have examined the issue of $CO_2$ lock-in from different perspectives.[27] For instance, Unruh's systemic interactions' perspective of carbon lock-in focuses on institutional and policy failures.[28] Unruh suggests that 'industrial economies have become locked into fossil fuel-based technological systems through a path-dependent process driven by technological and institutional increasing returns to scale'.[29] In this context, $CO_2$ lock-in 'arises through a combination of systematic forces that perpetuate fossil fuel-based infrastructures in spite of their known environmental externalities and the apparent existence of cost-neutral, or even cost-effective, remedies'.[30] Mattauch, Creutzig and Edenhofer take a different approach by focusing on the issue of market failure to explain $CO_2$ lock-in.[31] According to these commentators, one of the key barriers to a successful 'transformation to a low-carbon economy is the carbon lock-in', where 'fossil fuel-based ("dirty") technologies dominate the market', even in circumstances where their low-carbon or 'carbon-free ("clean") alternatives are dynamically more efficient'.[32] More recently, in a report on the imperative to transform the EU power sector, the European Environment Agency (EEA) has observed that in the energy system context, lock-in can be understood as 'mechanisms inhibiting the diffusion and adoption of carbon-saving technologies', resulting in a situation where 'the amount of fossil fuel capacity exceed the levels that correspond to [a State's] long-term de-carbonization objectives'.[33] More specifically, the EEA describes lock-in as 'a large (fossil fuel-based) technological overcapacity in the power sector, compared with its optimal configuration' that 'conveys a certain risk of path dependency';[34] where the 'self-perpetuating inertia created by large fossil fuel based energy infrastructure . . . inhibits public and private efforts to introduce alternative energy technologies' and 'energy efficiency measures designed to reduce GHG emissions'.[35] This accords with the IEA's discussion of electricity sector lock-in, namely the twin problems of dealing with the locked-in emissions generated by existing plants, while at the same time seeking to avert 'lock-in of emissions from future capacity'.[36] In this regard, the most recent global efforts to address the issues of rising GGEs and climate change are evidenced by the Paris Agreement.[37] This sets down new global aspirational aims of strengthening 'the global response to the threat of climate change', as well as significantly reducing the 'risks and impacts of climate change' by, *inter alia*, 'holding the increase in the global average temperature to well below 2°C above pre-industrial levels and pursuing efforts to limit the temperature increase to 1.5°C above pre-industrial levels'.[38] However, as the IEA has observed previously: 'Limiting global temperature rise to below 2°C will require a significant reduction in the use of coal in unabated power generation (i.e. from plants without CCS).'[39] More recently, the IEA has strengthened its position, suggesting that 'more action to reverse lock-in and avoid future lock-in' is required to satisfy the Paris Agreement's increased ambition to limit temperature increase to 'well below 2°C'.[40] One of the challenges in this regard is the extent to which the business-as-usual approach characterises a large part of current investment in new electricity generation infrastructure. For example, 60 per cent of the increase in electricity generation between 2001 and 2011 relied on 'lower efficiency subcritical technology'.[41] Action of this kind to date has effectively *locked in* the GHG emissions associated with electricity generation of this kind over the lifetime of the carbon-intensive infrastructure. This, the IEA has suggested, 'is not consistent with the IEA energy sector pathways to keep global temperature rise below 2°C';[42] that continuing to lock in 'high-emissions, long-lived infrastructure' is incompatible with 'keeping temperature rise well below 2° C'.[43] More so, 'if strong policy action is delayed until a new global climate treaty comes into effect after 2020', then the risk of further $CO_2$ lock-in is even higher.[44] While avoiding such outcomes requires deterrence of 'future installation of high carbon intensive technologies',[45] attention needs to be given to potentially perverse outcomes of decisions taken now.[46] For instance, while an initial decision to support switching from coal and oil to natural gas may be justified as being an important transitional step towards a low-carbon energy system, 'over-investment in natural gas in the short term could lead to lock-in of higher-emissions infrastructure that is incompatible with keeping temperature rise well below 2°C'.[47] There are also financial consequences of inadequate global reaction to this looming problem, with policy and practical response delays in shifting to a low-carbon pathway resulting in significant cost increases as time passes.[48] Addressing this issue requires, as the IEA explains, both a reduction in the power sector's 'heavy reliance on fossil fuels and the adoption of new, low-carbon generation and demand-side technologies'.[49] The integration of renewable energies into the network will help alleviate some of this problem and is already occurring in many regions such as the EU. Nevertheless, such changes are not without difficulties as such integration imposes 'new technical and economic challenges for electricity systems and for regulators'.[50] Overall, in terms of future trends in the power sector, the best that can be said is that many factors can impact on the extent to which requisite changes can be effected in practical settings. As the IEA comments:

> While change – and the need for change – is universal, the precise nature of these challenges, how the industry and public authorities are addressing them, and recent market trends vary considerably across countries and regions. This reflects the limited connectivity currently existing between (and, sometimes, within) national power systems, the diversity of utility ownership, policy approaches, big differences in market design, resource endowment and other local market characteristics.[51]

# 4 TRANSFORMING ELECTRICITY GENERATION TO MITIGATE $CO_2$ LOCK-IN: THE EU AND AUSTRALIAN EXPERIENCES

## 4.1 Challenging the Dominance of Fossil Fuels in the EU

### 4.1.1 EU energy and GGEs overview

EU energy markets continue to be dominated by fossil fuels (gas, solid fuels, and oil).[52] The natural flow-on of the underlying energy supply fuel mix is that energy products and services in the EU are likewise heavily dominated by fossil fuels. In 2015, the combined fossil fuels contribution to EU-28 total gross inland energy consumption was 72 per cent, albeit down from 81 per cent in 1995.[53] Combustible fuels (e.g., fossils fuels: natural gas, coal, and oil) made up just under half – 48.1 per cent – of the EU-28's net electricity generation fuel mix.[54] Encouragingly, solid fossil fuel consumption decreased in 2015 'for the third consecutive year'.[55] More importantly, renewable energies are gaining ground, albeit slowly. Non-fossil fuels' share of the EU energy fuel mix rose from 5 per cent in 1995 to 13 per cent in 2015,[56] as contributions from 'renewables, particularly biomass, wind and solar' continue to rise.[57] Relevantly, between 2005 and 2015, renewable energies' share of EU-28 net electricity generation fuel mix increased relative to coal, rising almost twofold from 13.3 to 25.3 per cent, with considerable increases in solar- and wind-generated electricity.[58]

Although the EU's total GGEs data for 2015 reveals a slight rise on 2014 levels,[59] the region's GGEs mitigation over last 25 years has been considerable: 2015 emissions were 23.6 per cent below 1990 levels.[60] The largest greenhouse gas in the EU is $CO_2$, which constitutes 81 per cent of the region's 2015 total GGEs (ex LULUCF – Land Use, Land Use Change and Forestry).[61] Even so, when compared to $CO_2$'s share of GGEs in 1990, 2015 levels of $CO_2$ emissions (ex LULUCF) are 22 per cent lower.[62] In terms of key sector contributions, the energy sector still prevails at the leading source of GGEs: 78 per cent of total GHG emissions in 2015 were generated by the combined energy sector.[63] In terms of individual energy sectors:

> 'Fuel combustion and fugitive emissions from fuels (without transport)' is responsible for 55% of EU-28 greenhouse gas emissions in 2015. In 1990 this source sector was even more dominant. Fuel combustion for transport (including international aviation) is the second most important source sector with 23% in 2015; it has increased its contribution significantly since 1990.[64]

Even so, since 1990, total GGEs from the EU energy sector have fallen by 23 per cent.[65]

### 4.1.2 Established, consistent policy foundations for a decarbonised EU energy market

As the European Commission (EC) has recently observed, there are a number of reasons for the overall sizeable reduction in the EU's GGEs discussed above. These include: (i) progressive decoupling of GDP and GGEs compared to 1990 levels;[66] (ii) growing shares in the use of renewables and less carbon-intensive fuels; (iii) reductions in 'energy-transformation losses' and energy efficiency improvements; (iv) structural changes in the economy; and (v) the economic recession; and 'policies (both EU and country-specific) . . . including key agricultural and environmental policies in the 1990s and climate and energy policies in the 2000s'.[67] The latter – discussed below – are particularly noteworthy as they demonstrate the EU's more holistic approach to mitigating GGEs, namely a comprehensive energy policy that embraces not only affordability, accessibility, and security, but also the environmental and social aspects of energy supply and use through a suite of measures. Relevantly, contemporary EU energy policy comprises three interrelated objectives: (i) ensure the uninterrupted physical availability of energy products and services on the market (secure supply – minimise energy security risks); (ii) at a price which is affordable for all consumers – private and industrial (competitive – price/cost), and (iii) at the same time, contribute to the EU's wider social and climate goals (sustainable – climate risks).[68] As part of the EU's approach to meeting these objectives, and its overarching goal of pursuing a low-carbon economy transformation pathway, climate and energy policy strategies have been formulated for different time frames: 2020, 2030, and 2050.[69] These include decisions to implement immediate and short-term collective targets for GGEs, renewable energy and energy efficiency. The temporal increases in the targets are intended to provide the EU with a stable policy framework to reduce the region's GGEs, increase the share of renewable energies in domestic consumption, and improve energy efficiency.[70] In turn, these targets are intended to provide 'investors more certainty' and confirm the 'EU's lead in these fields on a global scale'.[71] In addition to having a dedicated carbon pricing market, as evidenced though the maturing of the EU Emissions Trading System (EU ETS),[72] a range of EU legal measures support the various targets for GGEs mitigation, renewable energies, and energy efficiency.[73] The EC's recent inventory report on annual EU GGEs is instructive in terms of the region's progression along its decarbonisation pathway:

> Emissions from electricity and heat production decreased strongly since 1990. In addition to improved energy efficiency there has been a move towards less carbon intense fuels. Between 1990 and 2015, the use of solid and liquid

> fuels in thermal stations decreased strongly whereas natural gas consumption doubled, resulting in reduced $CO_2$ emissions per unit of fossil energy generated. Emissions in the residential sector also represented one of the largest reductions. Energy efficiency improvements from better insulation standards in buildings and a less carbon-intensive fuel mix can partly explain lower demand for space heating in the EU as a whole over the past 25 years. Since 1990 there has been a warming of the autumn/winter in Europe; although there is high regional variability. The very strong increase in the use of biomass for energy purposes has also contributed to lower GHG emissions in the EU.[74]

When it comes to the means for achieving the GGEs reductions mentioned above, it is remarkable to note that these have been achieved despite the major collapse in the carbon price in the EU ETS post 2008.[75] Even so, by reason of the weak carbon price, and the consequential lack of an adequate price signal, decarbonisation of the fuel mix for primary EU energy supply and electricity generation through this market mechanism has not eventuated to the extent expected to date.[76] Moreover, even with positive outcomes of the various energy and climate policies and implementation measures adopted by the EU to date, the ongoing heavy reliance on fossil fuels in electricity generation raises questions about the reality of the region meeting its 2030 targets, compared with 'the trends in the Energy Roadmap 2050, which deliver a 40% reduction in emissions from the energy sector by 2030'.[77] The EEA has recently considered the risk of $CO_2$ lock-in in this regard, concluding that

> under certain assumptions (in particular regarding the longevity of installed capacity), the EU power sector could evolve towards excessive fossil fuel capacity by 2030, compared with the optimal capacity levels in the Energy Roadmap 2050. The prolonged operation of inflexible, carbon-intensive power plants, along with the planned construction of new fossil fuel capacity, could translate into higher costs for decarbonising Europe's power sector by locking it in to a dependence on a high-carbon capacity, while simultaneously exposing owners and shareholders to the financial risk of capacity closures (potentially stranded assets).[78]

One of the looming obstacles to overcoming increased future $CO_2$ lock-in is the extent to which EU Member States' decisions regarding the 'adequacy of domestic generation' drive an increase in 'fossil fuel (and in particular solid fuel) overcapacity' and delays in 'decommissioning of fossil fuel capacity across Europe'.[79] More specifically, the EEA report suggests that:

> The levels of fossil fuel capacity decommissioning that the EU needs to achieve by 2030 to meet the Roadmap levels are equivalent to a 20–24% reduction in all fossil fuel capacity installed across the EU. For coal-fired capacity, they are

> equivalent to a 45% reduction in the installed capacity. In contrast, gas-fired capacity could increase by 6–11% of the capacity installed in 2014. Accordingly, between now and 2030 the EU needs to sustain an annual reduction in its fossil fuel capacity of between 1.75% and 2.0% to be in line with the Energy Roadmap 2050 decarbonisation levels. In contrast to this, in the absence of further incentives, such as a meaningful carbon price signal, it would realise an annual decrease of only 0.4–0.5% if the currently operational and planned units continued to operate until the end of their expected, longer, lifetimes. If it materialises, by 2050, this inertia in the power sector will have had considerable knock-on effects on the cost of reducing GHG emissions in the transport, residential and industrial sectors too.[80]

Consequently, the EEA's recommendation that there be 'rational and progressive decommissioning of fossil fuel capacity across the EU electricity sector' is certainly meritorious.[81] However, in light of the longevity of 'power sector infrastructure' and its resilience to change, 'new investment cycles in the sector can take decades to materialise and . . . the ensuing consequences are equally long lasting'.[82] In line with the Paris Agreement's support for non-State climate action across the board, mentioned previously, the EEA concludes that public and private sector organisations need to 'apply clear thinking to long-term planning, in order to avoid uncoordinated and more costly responses to the pressing decarbonisation challenges' characterising the energy sector. More specifically, the EEA calls for candid consideration of the evolution of the EU power sector, using the 'effective synergies between the climate and industrial emissions policies targeting' the sector to, *inter alia*: (i) develop a 'qualitatively different structure' for the short- and long-term future, which ensures that 'energy from renewable sources and energy efficiency progress in line with targets' and fossil fuel capacity decreases 'in line with decarbonisation objectives'; (ii) ensure 'both operators and regulators' are adequately equipped to manage the 'upcoming transition of the power sector and ongoing integration of the European energy market' as well as empowering 'organisations to come up with successful long-term business strategies in line with the EU's climate objectives'; and (iii) account for the regulatory risks 'posed by fossil fuel overcapacity and carbon infrastructure lock-in in the ongoing revision of the ETS and with regard to national initiatives that aim to establish capacity mechanisms (i.e. potential subsidies for extending the lifetime of capacity) in the power sector'.[83] Finally, the EEA concludes that

> the regular collection and dissemination of information on the actual and projected evolution of fossil fuel capacity across countries could be a useful complementary tool to enhance transparency and predictability for regulators and investors and to facilitate the cost-effective integration of cross-border

capacities. The ongoing discussions on governance tools under the Energy Union could, for example, lead to an agreement to provide such information as part of the integrated national energy and climate plans or the low carbon development strategies being advanced by countries.[84]

## 4.2 Supporting the Dominance of Fossil Fuels in Australia

### 4.2.1 Australian energy and GGEs overview

In Australia, the role of fossil fuels in the domestic energy sector largely mirrors the global snapshot: energy production, consumption, generation, and exports have continued their long-term upward trajectory.[85] Coal, natural gas, and oil continue to dominate the energy fuel mix in both domestic and export markets.[86] The natural flow-on from this is that energy products and services in Australia are likewise heavily dominated by fossil fuels. Consequently, even though dependency on coal in the national energy market has fallen since the turn of this century, this fossil fuel retains the largest share in the fuel source mix of the domestic electricity generation sector.[87] In 2016, coal's share in the electricity generation sector's fuel mix was 63.4 per cent.[88] This is still below its 2000–2001 peak, which exceeded 80 per cent.[89] Even so, reliance on coal has started to rise again.[90] The role of renewables in the national electricity market has grown considerably in the past decade. In 2015–16, renewable energy generation rose by 12 per cent, comprising 15 per cent of the fuel mix for this sector.[91] According to Australia's federal government, these changes have been 'driven by increases in hydro, wind and solar, which increased by 14, 6, and 24% respectively'.[92]

Like the EU, Australia's recently reported GGEs position appears to be a positive one: 2015 net emissions (including LULUCF) show a 9.3 per cent decrease on 1990 levels.[93] However, the changeable impact of LULUCF on the annual figures is considerable, varying between a net source or sink over the relevant time period (1990–2015).[94] By reason of the variable nature of LULUCF emissions, inclusion of this sector masks the true underlying, persistent sources of Australia's emissions. Indeed, a more meaningful appreciation of the trajectory in national emissions is provided by the data sets ex LULUCF. When that is considered, the situation is far from ideal, with Australia's net emissions (ex LULUCF) increasing by 27 per cent between 1990 and 2015.[95] Over the same 25-year period, emissions from the combined energy sectors and the stationary energy sector grew by around 43 per cent each.[96] In 2015, the combined energy sector (stationary energy, transport and fugitive emissions) constituted the largest source of GGEs, amounting to 78.7 per cent of Australia's total GGEs (ex LULUCF).[97] At the sub-sectorial level, between 1990 and

2015, GHG emissions increased across all sub-sectors: 'stationary energy (43.2%), transport (55.1%), fugitive emissions from fossil fuels (19.9%) and industrial processes and product use (24.0%)'.[98] In a similar vein, electricity generation emissions increased by 45.8 per cent between 1990 and 2015.[99]

### 4.2.2 Shifting, inconsistent policy foundations for a decarbonised Australian energy market

The historical reasons for fossil fuel's dominance of Australian energy markets are familiar: abundant resources, relative ease of availability, and low costs and prices. Reasons underpinning the ongoing rise in stationary energy emissions between 1990 and 2015 are said to result from increases in population, household incomes, and resource sector exports.[100] Relatively unambitious GGEs mitigation targets have also contributed to this situation. For example, targets for 2020 were set at 5 per cent reduction compared to 2000 levels,[101] and 28–30 per cent for 2030.[102] Both are significantly underwhelming when compared to the EU's targets outlined previously. Further reasons can also be advanced, including entrenched physical, legal, and regulatory systems that are underpinned by more than a decade of political apathy on the one hand, and deliberate attempts, on the other hand, to marginalise the renewable energy sector in favour of the fossil fuel incumbents. For instance, there is a notable absence of environmental considerations in Australian energy policy and laws, with various legislative mechanisms contained in separate, distinct laws. Thus, climate change risks and environmentally focused energy supply and use objectives are notably absent from Australia's existing national energy policy and legislative framework.[103] Moreover, Australia's embryonic steps towards establishing carbon pricing through a national ETS were short lived.[104] It was replaced with the Direct Action Plan, which includes a voluntary emissions reduction scheme that essentially pays polluters to reduce their emissions.[105] Likewise, an innovative, effective mechanism, aimed at promoting energy efficiency amongst the largest end-users – set down in the national Energy Efficiencies Opportunity Act – has also been abolished.[106] Thus, unlike in the EU, there is no energy-efficient target in place. In 2001, a renewable energy target was introduced to encourage the creation of a market for non-fossil fuels.[107] Renewable energies have been gaining ground on fossil fuels, but this market has suffered serious set-backs in recent years as a result of the actions of successive Conservative federal governments. Attempts to abolish this mechanism were thwarted, but even so, the magnitude of the existing renewable energy target has been curtailed with serious financial impacts.[108] More recently, the Australian government rejected the major finding of its own review into the national electricity market (the Finkel Review), namely to introduce a Clean Energy Target (CET) to help decarbonise Australia's electricity sector.[109] Similar to the EU energy and climate strategies, the Finkel Review's 2017 Final Report recommended that, *inter alia*, by 2020 the Australian government 'develop a whole-of-economy emissions reduction strategy for 2050'[110] that recognised the 'urgent need for a clear and early decision to implement an orderly transition' to a low-carbon future that 'includes an agreed emissions reduction trajectory, a credible and enduring emissions reduction mechanism and an obligation for generators to provide adequate notice of closure'.[111] The CET was the Finkel Review's preferred GGEs reduction mechanism, not only because it can 'deliver better price outcomes than business as usual' but also, it can be easily 'implemented within an already well understood and functioning framework', and has better price outcomes than the alternatives proposed.[112] In passing, it is worth noting that the Report drew attention to the strong collective views of the non-government actors, relevantly, a joint statement made by 'groups representing generators, networks, consumers, business and industry, unions, social services and environmental groups'.[113] This group of stakeholders collectively called for 'reforms of Australia's energy systems and markets to ensure reliability and affordability as we decarbonise the energy system' and the impacts on all Australians of the prevailing 'policy uncertainty, lack of coordination and unreformed markets', e.g., increasing costs, undermining of investment, and worsening reliability risks.[114] Moreover, the links between energy sector reform and GGEs reductions were expressed as follows: 'Energy has been a source of advantage for our industries and prosperity for our households. It should become so again even as Australian governments, businesses and communities deliver our national contribution to the global net zero emissions goal of the Paris Climate Agreement.'[115] In an interesting development, the current lack of transparent and progressive energy and climate policy – and requisite implementation mechanisms – has been noted by two of the key market institutions in the Australian Energy Market. The Australian Energy Market Commission (AEMC) recently suggested that 'without clear, national, coordinated policy objectives and credible mechanisms that reinforce one another both business and consumers find it difficult to invest'.[116] In a similar vein, the Australian Energy Regulator has observed: 'Uncertainty about governments' energy and climate change policies is affecting investor confidence. Outside of renewables, private investment in new plant has stalled while governments have announced plans to invest (or to explore investment) in gas, pumped hydro and energy storage.'[117] Even so, by comparison to the EU, the situation in Australia is one where, in addition to escalating domestic electricity and gas prices, national GHG emission levels reveal the inadequacy of national policy and legislative measures to effect meaningful mitigation to date. As the discussion above highlights, there is little support at the federal level for preventing, or at the very least minimising, electricity generation $CO_2$ lock-in, both now and into the future. As such, unless and until there is a change of federal government and/or a change in policy direction that takes the issue of mitigation energy-sector GGEs seriously, it is highly unlikely that Australia's electricity sector will avoid significant $CO_2$ lock-in.

## 5 CONCLUSIONS

In the context of shifting to a decarbonised energy future, the IEA's recent observations offer an insightful perspective on the need to lock in a low-, not high-carbon energy future:

> To achieve the necessary transformation of the world's energy system, the energy community must be persuaded that energy investments must be redirected so as to lock-in a widespread switch to low-carbon development that delivers economic growth and social development expectations, while simultaneously making deep cuts in emissions and strengthening global energy security. Significant climate action is already underway . . . But much of the transformational change needed in the energy sector to meet the 2°C climate goal has yet to take place. To bring this forth requires scaled-up national, regional and local action, guided by appropriate policies and standards, and mobilisation of both public and private finance for low-carbon energy supply and infrastructure.[118]

Interestingly, the 'amount of coal power capacity under development worldwide dropped dramatically from January 2016 to January 2017'.[119] The activities underlying these changes include reductions in construction projects (pre-construction, starts, ongoing, and completed) and an ongoing retirement of old power plants. As with the expansion between 2006 and 2016, changes in the domestic polices of China and India are at the heart of these changes.[120] However, as yet neither Australia, nor the EU, are fully equipped to avoid future electricity generation $CO_2$ lock-in. However, when contrasted with the EU's well-established approach to overcoming the energy sector's major role in climate change, Australia's adoption of relatively weak GGEs targets, together with an absence of credible federal energy *and* climate policy strategies and legal frameworks, is underwhelming at best. The EU's solid energy and climate policy and legal foundations make it highly likely that this region will manage the decarbonisation of its electricity generation sector considerably better than Australia will. Indeed, the current state of federal energy and climate affairs in Australia suggests a distinct lack of real interest in dealing with the current and future impacts of climate change. It will be unsurprising, therefore, if the comparison between the EU and Australia is even more stark in coming years, with the current Australian government's approach 'deterring investors and perpetuating the lock-in of high-carbon energy solutions'.[121]

## NOTES

1. In some cases, there is no other end use energy supply option, e.g., electrical appliances.
2. IEA, *$CO_2$ Emissions from Fuel Combustion: Overview (2017 edition)* (OECD/IEA 2017) 4–5.
3. IEA, *Energy, Climate Change and Environment: 2016 Insights* (OECD/IEA 2016) 28. See also B Naughton, 'Emissions Pricing, "Complementary Policies" and "Direct Action" in the Australian Electricity Supply Sector: Some Conditions for Cost-Effectiveness' (2013) 32(4) *Economic Papers* 440, 445, where the problem is examined in terms of locking in 'to existing or imminently installed coal-fired generating capacity and other high $CO_2$-intensive generating technologies'.
4. IEA, *Key World Energy Statistics* (*KWES*) (OECD/IEA 2017) 6, 34.
5. Ibid, 30.
6. Ibid.
7. Ibid.
8. Ibid, 6, 34 and 30.
9. Ibid, 8 and 38.
10. Ibid.
11. Ibid, 32.
12. See e.g., IEA, 'China 13th Renewable Energy Development Five Year Plan (2016–2020)' <https://www.iea.org/policiesandmeasures/pams/china/name-161254-en.php>.
13. IEA, *KWES*, above n 4, 54.
14. Ibid.
15. IEA, *$CO_2$ Emissions from Fuel Combustion*, above n 2, 5.
16. IEA, *KWES*, above n 4, 55.
17. Ibid.
18. Ibid; IEA, *$CO_2$ Emissions from Fuel Combustion,* above n 2, 6.
19. See UNFCCC official website <http://unfccc.int/2860.php> for international treaties and discussions of same.
20. The Paris Agreement (adopted 12 December 2015; entered into force 4 November 2016) <http://unfccc.int/paris_agreement/items/ 9485.php>. Currently, much of the Paris Agreement's operative effect is under negotiation and discussions. See UNFCCC official website <http://unfccc.int/2860.php> for further detailed discussion on the implementation work plan under way in preparation for the next Conference of Parties – the COP24 – in Poland, 2018.
21. IEA, *Energy Policies of IEA Countries: EU 2014 Review* (OECD/IEA 2014) 202.
22. IEA, *Energy, Climate Change and Environment: 2014 Insights* (OECD/IEA 2014) 15.
23. IEA, *Energy and Air Pollution, World Energy Outlook Special Report* (OECD/IEA 2016) 43.
24. C Shearer, N Ghio, L Myllyvirta, A Yu, and T Nac, *Boom and Bust 2017: Tracking the Global Coal Plant Pipeline* (Coalswarm, Sierra Club, and Greenpeace, 2017) 3.
25. IEA, *World Energy Outlook 2014* (OECD/IEA 2015) 204, 204.
26. Ibid 202.

27. See e.g., European Environment Agency (EEA), *Transforming the EU Power Sector: Avoiding a Carbon Lock-in*, EEA Report No 22/2016 (EEA, 2016); IEA, *2016 Insights*, above n 3; N Johnson, V Krey, DL McCollum, S Rao, K Riahi and J Rogelj, 'Stranded on a Low-carbon Planet: Implications of Climate Policy for the Phase-out of Coal-based Power Plants' (2015) 90(A) *Technological Forecasting and Social Change* 89; A Klitkou, S Bolwig, T Hansen, N Wessberg, 'The Role of Lock-in Mechanisms in Transition Processes: The Case of Energy for Road Transport' (2015) 16 *Environmental Innovation and Societal Transitions* 22; IEA, *Energy and Climate Change* (OECD/IEA 2015); B Naughton, above n 3; L Mattauch, F Creutzig, O Edenhofer, 'Avoiding carbon lock-in: Policy options for advancing structural change' (2012) 50 *Economic Modelling* 49; N Frantzeskaki and D Loorbach, 'Avoiding carbon lock-in: Policy Options for Advancing Structural Change' (2012) 50 *Economic Modelling* 49; N Frantzeskaki *et al*, 'Towards Governing Infrasystem Transitions: Reinforcing Lock-in or Facilitating Change?' (2010) 77(8) *Technological Forecasting and Social Change* 1292; GC Unruh, 'Understanding Carbon Lock-in' (2000) 28 *Energy Policy* 817.
28. Unruh, above n 27.
29. Ibid.
30. Ibid, 818, where the author coins the term *Techno-Institutional Complex* to explain the technology–institutional relationship, arguing that this 'arises because large technological systems, like electricity generation, distribution and end use, cannot be fully understood as a set of discrete technological artefacts but have to be seen as complex systems of technologies embedded in a powerful conditioning social context of public and private institutions'.
31. Mattauch et al, above n 27.
32. Ibid; Unruh, above n 27.
33. EEA, above n 27, 7.
34. Ibid, 5.
35. Ibid, 61.
36. IEA, *2016 Insights*, above n 3, 28. See also Naughton, above n 3, 445, where the problem is examined in term of locking in 'to existing or imminently installed coal-fired generating capacity and other high $CO_2$-intensive generating technologies'.
37. The Paris Agreement (adopted 12 December 2015; entered into force 4 November 2016) available online <http://unfccc.int/paris_agreement/items/9485.php>.
38. Ibid, Article 2.
39. IEA, *2016 Insights*, above n 3, 28.
40. Ibid.
41. IEA, *EU 2014 Insights*, above n 22, 15.
42. Ibid.
43. Ibid. See also IEA, *2016 Insights,* above n 3, 104.
44. Ibid.
45. Naughton, above n 3, 445.
46. See e.g., EEA, above n 27, where the unintended consequences of the EU's 2010 Directive on industrial emissions are discussed.
47. IEA, *2016 Insights*, above n 3.
48. Ibid.
49. IEA, *EU 2014 Review*, above n 21, 202.
50. Ibid.
51. Ibid.
52. EC, *EU Energy in Figures: Statistical Pocketbook 2017* (Publications Office of the EU, 2017). See also EC, *Annual EU GHG Inventory 1990–2015 and Inventory Report 2017* (EEA Report, Submission to the UNFCCC Secretariat, 2017).
53. EC, *EU Energy in Figures*, above n 52, 22.
54. Eurostat, 'Electricity production, consumption and market overview' (June 2017) <http://ec.europa.eu/eurostat/statistics-explained/index.php/Category:Electricity>.

This rose to 48.6 per cent in 2016: see Eurostat, 'Electricity generation statistics – first results' (May 2017) <http://ec.europa.eu/eurostat/statistics-explained/index.php/Category:Electricity>.
55. EC, *Annual EU GHG Inventory*, above n 52, 69.
56. Ibid.
57. Ibid.
58. Eurostat, 'Electricity production, consumption and market overview', above n 54, where it is noted that 'the proportion of net electricity generated from solar and wind increased greatly: from less than 0.1% in 2005 to 3.5% in 2015 for solar power and from 2.2% in 2005 to 9.7% in 2015 for wind turbines'.
59. EC, *Annual EU GHG Inventory*, above n 52, iii.
60. Ibid.
61. Ibid, 65.
62. Ibid.
63. Ibid, 80.
64. Eurostat, 'Greenhouse gas emission statistics – emission inventories' (EC, June 2017) <http://ec.europa.eu/eurostat/statistics-explained/index.php/Greenhouse_gas_emission_statistics_-_emission_inventories>.
65. EC, *Annual EU GHG Inventory*, above n 52, 80.
66. Ibid, where it is pointed out that between 1990 and 2015, GDP increased by around 50 per cent, while GGEs decreased by around 24 per cent.
67. Ibid, iv–vi.
68. See e.g., EC, 'An Energy Policy For Europe', Communication from the Commission (10 January 2007) COM (2007) 1 final; EC, 'Second Strategic Energy Review, An EU Energy Security and Solidarity Action Plan', Communication from the Commission (13 November 2008) COM (2008) 781 final; EC, 'Energy 2020 A strategy for competitive, sustainable and secure energy', Communication from the Commission (10 November 2010) COM (2010) 639 final; EC, 'A Framework Strategy for a Resilient Energy Union with a Forward-Looking Climate Change Policy', COM (2015) 80 final, Brussels, 2015.
69. See e.g., EC, 'An Energy Policy For Europe', Communication from the Commission (10 January 2007) COM (2007) 1 final; 'Second Strategic Energy Review, An EU Energy Security and Solidarity Action Plan', Communication from the Commission (13 November 2008) COM (2008) 781 final; 'Energy 2020 A strategy for competitive, sustainable and secure energy', Communication from the Commission (10 November 2010) COM (2010) 639 final; 'A policy framework for climate and energy in the period from 2020 to 2030', Communication from the Commission (22 January 2014) COM (2014) 15 final, where targets include 40 per cent GGEs mitigation, and 27 per cent increase in share of renewable energy and energy efficiencies by 2030; 'A Roadmap for moving to a competitive low carbon economy in 2050', Communication from the Commission (8 March 2011) COM (2011) 0112 final; 'Energy Roadmap 2050', Communication from the Commission (12 December 2011) COM (2011) 0885 final, which recognise, inter alia, the ambitious target for GHG emissions reductions of 80–95 per cent by 2050.
70. See also more recently, EC, 'A Framework Strategy for a Resilient Energy Union with a Forward-Looking Climate Change Policy', COM (2015) 80 final, Brussels, 2015.
71. DG Energy <https://ec.europa.eu/energy/>.
72. Directive 2009/29/EC of the European Parliament and of the Council amending Directive 2003/87/EC so as to improve and extend the greenhouse gas emission allowance trading scheme of the Community (OJ L 140, 5.6.2009); Directive 2008/101/EC of the European Parliament and of the Council amending Directive 2003/87/EC so as to include aviation activities in the scheme for greenhouse gas emission allowance trading within the Community (OJ L 8, 13.1.2009); Directive 2004/101/EC of the European Parliament and of the Council amending Directive 2003/87/EC establishing a scheme for greenhouse gas emission allowance trading within the Community, in respect

of the Kyoto Protocol's project mechanisms (OJ L 338, 13.11.2004); Consolidated version of Directive 2003/87/EC of the European Parliament and of the Council establishing a scheme for greenhouse gas emission allowance trading within the Community and amending Council Directive 96/61/EC (OJ L 275, 25.10.2003).

73. Directive 2009/28/EC of the European Parliament and of the Council of 23 April 2009 on the promotion of the use of energy from renewable sources (OJ L 140, 5.6.2009); Decision No 406/2009/EC of the European Parliament and of the Council of 23 April 2009 on the effort of Member States to reduce their greenhouse gas emissions to meet the Community's greenhouse gas emission reduction commitments up to 2020 (OJ L 140, 5.6.2009); Directive 2009/31/EC of the European Parliament and of the Council of 23 April 2009 on the geological storage of carbon dioxide (OJ L 140, 5.6.2009); Directive 2003/30/EC of the European Parliament and of the Council of 8 May 2003 on the promotion of the use of biofuels or other renewable fuels for transport (OJ L 123, 17.5.2003); Regulation (EC) No 443/2009 of the European Parliament and of the Council of 23 April 2009 setting emission performance standards for new passenger cars as part of the Community's integrated approach to reduce $CO_2$ emissions from light-duty vehicles (OJ L 140, 5 June 2009); Commission Regulation (EU) No 651/2014 and the provision of state aid for energy-environment matters.
74. *Annual EU GHG Inventory*, above n 52, iv; further discussion at 69 ff.
75. IEA, *EU 2014 Review*, above n 21, 57–8; See also EEA, *Trends and Projections in Europe 2014: Tracking Progress Towards Europe's Climate and Energy Targets for 2020*, EEA Report No 6/2014.
76. See e.g., EEA, above n 27.
77. Ibid, 61.
78. Ibid.
79. Ibid.
80. Ibid, 66.
81. Ibid, 68.
82. Ibid.
83. Ibid, 68, where the importance of 'avoiding costly technological retrofits in the power sector' is noted, 'to reduce air pollution emissions and comply with the [Industrial Emissions Directive] . . . and reduce the risk of future stranded assets, especially among the most carbon-intensive and inflexible baseload plants that are not able to support the integration of an intermittent power supply from renewable sources'.
84. Ibid, 68.
85. For further detailed information see Department of the Environment and Energy (DEE), *Australian Energy Update 2017* (Canberra, August 2017), esp. 12, where *energy consumption* is defined as 'the amount of energy used in the Australian economy. It is equal to indigenous production plus imports minus exports (and changes in stocks). It includes energy consumed in energy conversion activities (such as electricity generation and petroleum refining), but excludes derived or secondary fuels (such as electricity and refined oil products) produced domestically to avoid double counting of energy.'
86. For further detailed information see DEE, *Australian Energy Statistics* <https://www.energy.gov.au/government-priorities/energy-data/australian-energy-statistics>.
87. DEE, *Australian Energy Update 2017*, above n 85.
88. Ibid, 24.
89. Ibid.
90. Ibid, 9 and 24–5.
91. Ibid, 26.
92. Ibid, 26–7, where the observation is made that these changes are 'mainly attributable to dam levels recovering after dry spells in the previous year, and additional investment in renewable capacity supported by the Renewable Energy Target'.
93. Australian Government, *National Inventory Report 2015 Volume 1* (Commonwealth of Australia 2017) 32.

94. Ibid.
95. Ibid, 32.
96. Ibid, xi and 32 ff. These values are based on the data set out in *Table ES.01 Australia's net GGEs by sector under the UNFCCC, 1990, 2005, 2014 and 2015,* xi and discussions 32 ff.
97. Ibid, 32.
98. Ibid, 34.
99. Ibid, 39.
100. Ibid, where it is also pointed out that increases in the emissions intensity of delivered electricity are the reason for more recent rises in electricity generation emissions. It is also noted that: 'The growth in transport emissions is attributed to 'the number of passenger vehicles, along with an increase in diesel consumption in heavy vehicles'. The increase in fugitive emissions is said to result from 'increased production from open cut coal mines and increased gas production'.
101. See further discussion of 2020 and 2030 targets in CCA, *Reducing Australia's GHG Emissions: Targets and Progress Review – Final Report* (Commonwealth of Australia, 2014) esp. Ch 9.
102. Australian Government, 'Australia's 2030 Emission Reduction Target' (Commonwealth of Australia, 2015) <http:// www.environment.gov.au/climate-change/government/australias-emissions-reduction-target>.
103. This has been discussed extensively by this author in N Ison, J Usher, R Cantley-Smith, S Harris, and C Dunstan, *Report Card on the National Electricity Market* (NEM) (Australian Energy Consumers, 2011); R Cantley-Smith, 'A Changing Legal Environment for the National Electricity Market' in W Gumley and T Daya-Winterbottom (eds), *Climate Change Law: Comparative, Contractual and Regulatory Considerations* (Lawbook, 2009).
104. The carbon pricing scheme (carbon tax/ETS) introduced in the Clean Energy Act 2011 (Cth) was repealed in 2014.
105. Carbon Credits (Carbon Farming Initiative) Act 2011 (Cth); Carbon Credits (Carbon Farming Initiative) Regulations 2011; Carbon Credits (Carbon Farming Initiative) Rule 2015.
106. This was discussed in an earlier GCET publication: R Cantley-Smith, 'Demanding More: The Role of Demand Management and Improved End Use Efficiency in Australian Electricity Markets' in J Cottrell, JE Milne, H Ashiabor, LA Kreiser, and K Deketelaere (eds), *Critical Issues of Environmental Taxation*, Vol VI (Oxford University Press, 2009).
107. This mechanism was discussed in an earlier GCET publication: R Cantley-Smith, 'Environmental Regulation of Australian Energy Markets: Are Mandatory Renewable Targets Effective in Reducing Greenhouse Gases?' in K Deketelaere et al (eds), *Critical Issues of Environmental Taxation*, Vol IV (Oxford University Press, 2007).
108. For example, in the year following this change, investment in the renewable energy sector fell by around 80 per cent.
109. A Finkel, K Moses, C Munro, T Effeney and M O'Kane, *Independent Review into the Future Security of the National Electricity Market: Blueprint for the Future* (Commonwealth of Australia, 2017).
110. Ibid, 23, Rec 3.1.
111. Ibid, Rec 3.2.
112. Ibid.
113. Ibid, 31, referring to AI Group, 'Joint Statement: Energy reform is urgent to avert systemic crises' (13 December 2016) <https://www.aigroup.com.au/policy-and-research/mediacentre/releases/Joint-Statement-Energy-Reform-13Dec/>.
114. AI Group, above n 113.
115. Ibid.
116. This mechanism was discussed in an earlier GCET publication, Cantley-Smith, above n 107.

117. AER, *State of the Energy Market 2017* (ACCC, 2017).
118. IEA, *Energy and Climate Change*, above n 27, 132.
119. Shearer et al, above n 24.
120. Ibid, 3–4, where the authors note: 'The main cause of the shrinkage in the coal plant pipeline was the imposition of unprecedented and far-reaching restrictive measures by China's central government.'
121. IEA, *Energy and Climate Change*, above n 27, 134.

# 15. An overview of zero emission credits for nuclear power plants in the United States

**Hans Sprohge and Larry Kreiser**

## 1 INTRODUCTION

Zero emission credits (ZEC) are subsidies only to ageing nuclear power plants unable to compete in the marketplace due to low natural gas prices. State legislatures in Illinois and New York have approved up to $10 billion in ZEC subsidies over the next decade.[1] Many other states are considering enacting similar subsidies. If every nuclear reactor in the Northeast and Mid-Atlantic States were to receive subsidies at the same level as those in Illinois and New York, the total subsidies would amount to $3.9 billion annually.[2]

Lawmakers contemplating enacting ZECs for ageing nuclear power plants unable to compete in the marketplace face contradictory information. On the one hand, countless sources of information provide arguments in favor of enacting ZECs. On the other hand, many sources of information maintain that the arguments in favor of enacting ZECs are based on the premise that human activity increases atmospheric carbon dioxide ($CO_2$) causing global warming resulting in disastrous environmental consequences. These sources argue that this premise is false since it is based on misleading information. So, lawmakers are in a dilemma which could lead to wasting billions of dollars to resolve a nonexistent problem. Lawmakers cannot rely on contradictory conclusions when deciding on whether or not to enact ZECs even if they are published by credentialed scientists in peer-reviewed journals. Global warming or no global warming is a false dichotomy.

What lawmakers need to consider are the risks that enacting ZECs may exacerbate, by extending the operating life of nuclear power plants, nuclear meltdowns, cyberattacks, and terrorist attacks on nuclear waste. In light of the risks exacerbated by extending the life of ageing nuclear power plants, enacting ZECs is irresponsible and shortsighted.

## 2 ARGUMENTS IN FAVOR OF ENACTING ZECs

Some arguments that have been raised in support of ZECs to keep ageing nuclear power plants operational are as follows: nuclear power plants that cease to operate will be replaced by fossil-fuel plants that will emit million of tons of extra carbon dioxide every year and other air pollutants such as nitrogen oxide.[3] These greenhouse gases will increase public health costs by billions of dollars.[4] The lost zero emissions capacity from nuclear power plants could not be replaced for many years by other zero emissions fuel sources. In the interim, the increase in carbon dioxide emissions may make compliance with the Environmental Protection Agency's (EPA) Clean Power Plan impossible.[5] The loss of nuclear plants would also make the power grid less secure because of supply interruptions of the kind seen in the recent polar vortex episode and worsen grid instability because of a growing share of intermittent renewable power.[6] In addition, keeping ageing nuclear plants operational preserves good paying jobs. A 2014 study found that nuclear reactors generate $40 billion to $50 billion in annual electricity sales while employing more than 100,000 workers.[7] These and many other arguments in support of ZECs are based on the premise that human emissions of $CO_2$ into the atmosphere are the primary cause of global warming with catastrophic environmental consequences.

## 3 ARGUMENTS IN FAVOR OF ENACTING ZECs ARE BASED ON A MISLEADING PREMISE

The validity of the global warming premise is thrown into doubt by misleading information. Misleading information is created by a variety of studies purporting to demonstrate a 97 percent consensus among climate scientists on global warming. Two frequently cited studies are authored by Naomi Oreskes and John Cook. In her article,[8] Oreskes asserts that 'without substantial disagreement, scientists find human activities are heating the earth's surface.' Oreskes states that her examination of 928 abstracts in the Institute for Scientific Information database containing the phrase 'climate change' of articles published in scientific journals between 1993 and 2003 show that 75 percent support the view that human activities are responsible for most of the observed global warming over the previous 50 years while none directly dissented from this position. Her study is questionable for the following reasons: Her definition of consensus covers 'man-made' influences but leaves out 'dangerous' or some synonym indicating disastrous consequences. Her list excludes scores of studies that show that 'global temperatures were similar or even higher during the Holocene Climate Optimum and the Medieval Warm Period when atmospheric $CO_2$ levels were much lower than today.' Oreskes' study also does not take into account that abstracts of academic papers often contain unsubstantiated claims.

The Cook study examined 11,944 peer-reviewed scientific literature abstracts from 1991–2011 matching the topics 'global climate change' or 'global warming.'[9] The results were as follows:

- 66.4 percent of abstracts expressed no position on anthropogenic global warming (AGW);
- 32.6 percent endorsed AGW;
- 0.7 percent rejected AGW;
- 0.3 percent were uncertain about the cause of global warming.

Of the abstracts taking a position on AGW, 97.1 percent endorsed the position that humans cause global warming. In a second phase of this study, when authors rated their own papers, a smaller percentage of self-rated papers expressed no position on AGW (35.5 percent). Of the self-rated papers taking a position on AGW, 97.2 percent endorsed the position that humans contribute to global warming. An examination of Cook's own tables shows that only 1.6 percent of the surveyed abstracts clearly say that humans are the primary cause of global warming.[10] The 97.1 percent figure includes papers that merely claim or imply that some amount of global warming is attributable to human activity. Cook created a category called 'explicit endorsement without quantification' – that is, papers that did not say whether 1 percent or 50 percent or 100 percent of the warming was caused by humans.[11] He also created a category called 'implicit endorsement' for papers that implied that there is some human-activity global warming but do not quantify it. 'Explicit endorsement without quantification' and 'implicit endorsement' are somewhat misleading labels because they are a far cry from the notion that humankind is the primary source of disastrous global warming.

Historical temperature data for the continental United States have been adjusted by the National Aeronautics and Space Administration (NASA) to show US temperatures trending upward. The charts in Figure 15.1, published by NASA on its website in 1999, show an inconsistency between US and global temperatures.[12]

Chart (a) in Figure 15.1 shows that the highest temperatures in the United States occurred in the 1930s followed by a downward cooling trend to the year 2000. The website even contains the following statements: 'Empirical evidence does not lend much support to the notion that the

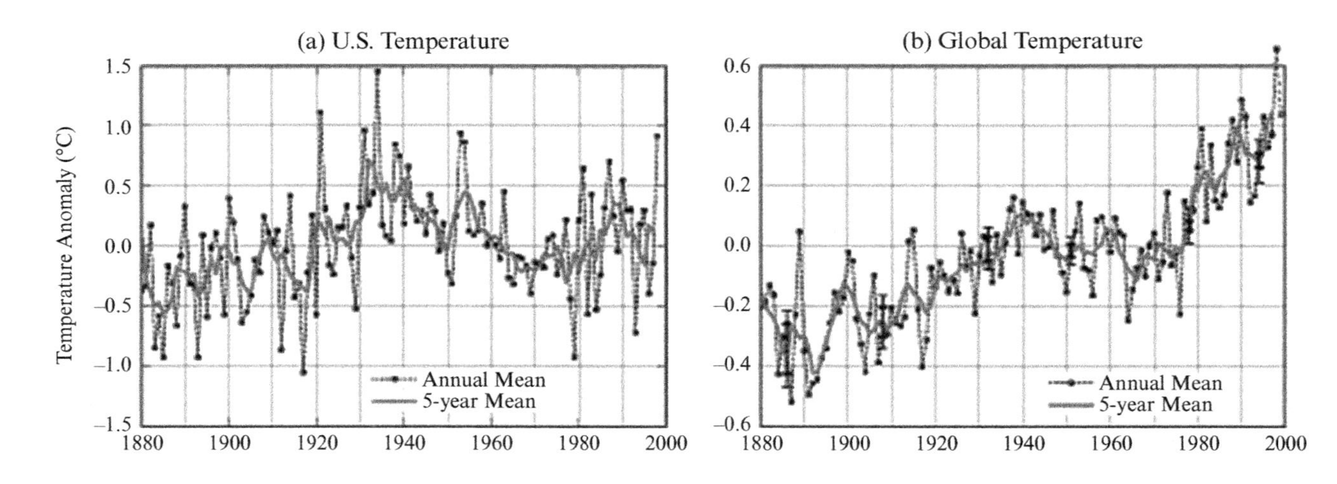


*Source:* James Hansen, Reto Ruedy, Jay Glascoe and Makiko Sato. 'Whither U.S. Climate?' *Science Briefs*, NASA, August 1999, www.giss.nasa.gov/research/briefs/hansen_07/.

*Figure 15.1 US temperatures and global temperatures (1880–2000)*

climate is headed precipitately toward more extreme heat and drought' and '. . . in the U.S. there has been little temperature change in the past 50 years, the time of rapidly increasing greenhouse gases – in fact, there was a slight cooling throughout much of the country.' Chart (b) shows global warming over the same time period.

Right after the year 2000, NASA adjusted the data in Chart (a) to be consistent with Chart (b). It did so by decreasing the severe heat and droughts experienced in the 1930s to make them cooler and adjusted upward the temperature data after the 1970s to make them warmer. As Figure 15.2 shows, these adjustments converted a long-term cooling trend since 1930 into a warming trend.[13]

Although adjustments are necessary to account for 'biases' in the data, a recent study concludes that what makes the NASA adjustments questionable is that they are inconsistent with published and credible US and other temperature data.[14] This same study shows that climate data adjustment is not limited to NASA and the National Oceanic and Atmospheric Administration (NOAA). The University of East Anglia's Climatic

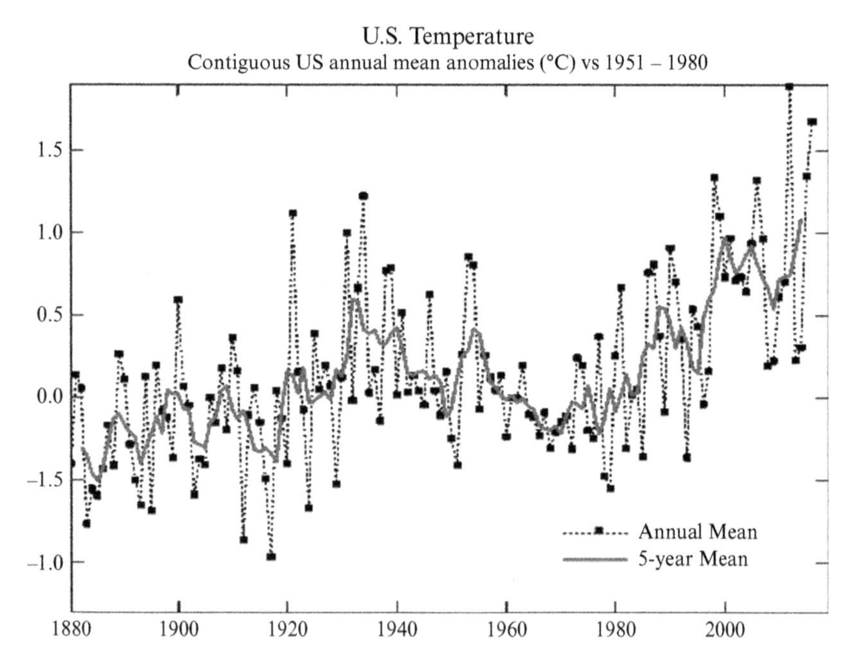


*Source:* 'Annual Mean Temperature Change in the United States.' GISS Surface Temperature Analysis, NASA, 13 October 2016, data.giss.nasa.gov/gistemp/graphs_v3/.

*Figure 15.2 US temperatures adjusted by NASA*

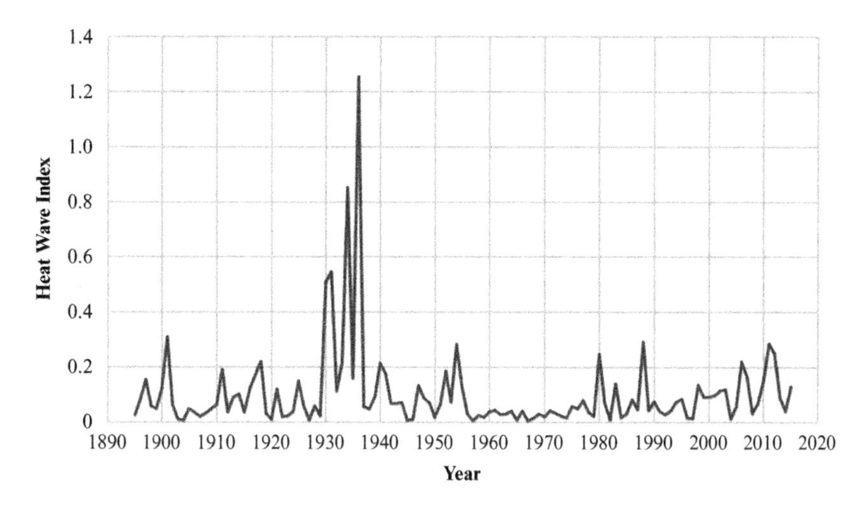


*Source:* 'Climate Change Indicators: High and Low Temperatures.' *Climate Change Indicators*, Environmental Protection Agency, 17 Dec. 2016, www.epa.gov/climate-indicators/climate-change-indicators-high-and-low-temperatures.

*Figure 15.3 EPA heat wave index*

Research Unit (CRU)/Hadley Center and the NOAA also adjusted climate data.[15] Other entities adjusting temperature data include the New Zealand government's National Institute of Water and Atmospheric Research,[16] Wikipedia,[17] and the United Nations Intergovernmental Panel on Climate Change.[18]

The EPA's 'Heat Wave Index' data supports the contention that the United States was far hotter in the 1930s than it is today. Figure 15.3, published on the EPA's website, shows that recent heat waves are far smaller and less severe than those of the 1930s.[19] The heat waves of the last few years were no worse than those of the early 1900s or 1950s.

## 4 THE LAWMAKERS' DILEMMA

Lawmakers who want to enact ZEC legislation to protect the environment are caught on the horns of a dilemma because of conflicting information available to them. On the one hand, many considerations supporting ZEC legislation are put forth from a variety of creditable sources. These considerations are based on the contention that global warming is caused by increases in $CO_2$ in the atmosphere from human activity and that global warming is disastrous for the environment. On the other hand, doubts about this contention are countered by misleading information. Many of these doubts are put forth in a variety of creditable sources such as peer-reviewed professional journals. So, who are the lawmakers supposed to believe: those who say there is global warming or those who say there is no global warming?

With respect to enacting ZEC legislation, lawmakers do not have to resolve the question of whether global warming is caused by human activity increasing $CO_2$ in the atmosphere. Instead, lawmakers can look at how enacting ZECs exacerbates environmental risks by extending the operating life of nuclear power plants. These environmental risks include nuclear meltdowns, cyber attacks on nuclear facilities, and terrorist attacks on nuclear waste.

## 5 NUCLEAR MELTDOWNS

After the 1979 reactor core meltdown at the Three Mile Island nuclear plant in Pennsylvania, the US Nuclear Regulatory Commission (NRC) established a safety policy that estimates the probability of a nuclear reactor meltdown, on a nuclear plant licensed to operate for 40 years, to be no more than once every 10,000 years of reactor operation.[20] Since then, the NRC has extended the operating licenses of more than 75 percent of 100 US nuclear reactors by 20 years and is considering extending the licenses for an additional 20 years up to 80 years.[21] The ageing of nuclear power plants is a risk factor currently ignored by the NRC. As nuclear power plants age, their structures, systems, and components degrade. Steel reinforcements grow weak and rusty, concrete crumbles, paint peels, and crud accumulates.[22] When the NRC began renewing licenses, it did not revisit its safety goal. The *Bulletin of the Atomic Scientists* estimates the chance of one nuclear reactor meltdown among a fleet of 100 reactors operating within the NRC's safety goal for 40 years is nearly 33 percent. The chance of a meltdown from the fleet operating for 60 years rises to nearly 45 percent. The meltdown risk from the fleet operating for 80 years is nearly 55 percent.[23]

## 6 CYBER ATTACKS ON NUCLEAR FACILITIES

Ageing nuclear power plants are potential targets for cybersecurity attacks.[24] Hackers have succeeded in penetrating the firewalls and digital protections of administration information at Wolf Creek nuclear power plant near Burlington, Kansas. Computer systems controlling critical nuclear operations at Wolf Creek were never compromised during the breach because they are separate from business computer networks by physical distance and hardware. Security experts say that a successful cyber attack on a nuclear plant by terrorists using new technology is all but inevitable with devastating results.[25] Hundreds of square miles could be contaminated with long-lived radioactive material, forced resettlement of millions of people may be required, and damage could run in the trillions of dollars.[26]

There is also another way nuclear plants are vulnerable to cyber attacks: the electrical grid. If hackers shut down the grid for an extended period of time, the loss of power to the safety systems that cool the nuclear fuel and regulate the reactor could fail despite backup emergency generators and batteries.

## 7 TERRORIST ATTACKS ON NUCLEAR WASTE

Waste from nuclear power plants consists of spent fuel rods and some fuel processing waste. The radioactivity from this waste will be lethal for 250,000 years.[27] The Great Pyramid of Giza was built about 3,200 BC which is only a little more than 5,000 years ago.[28] The preferred solution to nuclear waste in almost all countries is a geological repository.[29] Currently, the United States does not have a location for permanent storage of spent fuel.[30] When a geological repository is established in the United States, thousands of shipments of nuclear waste will be transported on railways and highways for many years. Terrorists could hijack such shipments or attack them with mortars or rockets, threatening the lives of millions of people.[31] Aside from terrorist attacks on shipments of nuclear waste, there is also no way to predict repository behavior over geologic time. There is no way to predict changes in weather, volcanic activity, plate-tectonic movement, etc. tens of thousands of years into the future. Furthermore, communicating the dangers of buried nuclear waste for tens of thousands of years into the future borders on science fiction.[32]

Because no geological repositories exist, thousands of tons of highly radioactive used reactor rods are currently stored in large pools of water.[33] The spent fuel rods are kept from overheating by the water in these pools, which absorbs radiation. This waste could ignite if exposed to air, causing a catastrophic fire that could be worse than a reactor meltdown.[34] Many spent fuel pools are above ground and are protected by corrugated buildings. Terrorists could displace or evaporate enough water to leave the rods exposed to air by breaching the walls of the protective buildings with a plane crash or even a large truck bomb. A report released in April 2006 by the National Academy of Sciences (NAS) found that 'successful terrorist attacks on spent fuel pools, though difficult, are possible.' 'If an attack leads to a propagating zirconium cladding fire, it could result in the release of large amounts of radioactive material.'[35]

## 8 CONCLUSIONS

Some states are considering legislation that, if enacted, would create billions of dollars in ZEC subsidies to ageing nuclear power plants similar to legislation already passed in Illinois and New York. In looking for a rational basis to decide whether or not to enact such legislation, lawmakers are faced with a staggering amount of contradictory information from creditable sources. Some sources maintain that the Earth will be exposed to dire consequences by global warming caused by human activity releasing $CO_2$ into the atmosphere. Other sources maintain that global warming from human activity does not exist because there is no scientific consensus and some measurements of the Earth's temperature show global cooling instead of global warming. Which position on global warming is right should be irrelevant to lawmakers when considering ZEC legislation because of other environmental risks posed by nuclear power plants. These risks include nuclear meltdowns, cyber attacks on nuclear facilities, and terrorist attacks on nuclear waste. In summary, enacting ZEC legislation for ageing nuclear power plants in order to extend their life is shortsighted in its approach and entails significant environmental risk.

## NOTES

1. Perry, Mark J. 'Why a Multibillion-Dollar Bailout for Nuclear Plants Would Be a Colossal Blunder.' *OPINION Contributors*, Washington Examiner, 16 May 2017, www.washingtonexaminer.com/why-a-multibillion-dollar-bailout-for-nuclear-plants-would-be-a-colossal-blunder/article/2623212.
2. Ibid.
3. Bayles, Jessica. 'NY Creates New Emissions Credit for Nuclear Plants.' *Energy Business Law*, McDermott Will & Emery, 20 Sept. 2016, www.energybusinesslaw.com/2016/09/articles/environmental/ny-creates-new-emissions-credit-for-nuclear-plants/.
4. London, Herbert. 'Cuomo's nuclear bailout a bad deal for state.' *timesunion*, Hearst, 19 Nov. 2016, www.timesunion.com/tuplus-opinion/article/Cuomo-s-nuclear-bailout-a-bad-deal-for-state-10625370.php.
5. Brand, Stewart, et al. 'New York Public Service Commission Letter.' *Climate Scientists and Conservation Leaders Urge New York Public Service Commission to Save Nuclear Power*, Environmental Progress, 14 July 2016, www.environmentalprogress.org/new-york-public-service-commission-letter.

6. Ibid.
7. Joyce, Stephen. 'Nuclear Subsidies Push Seen Spreading to New States.' *News*, Bloomberg BNA, 19 July 2017, www.bna.com/nuclear-subsidies-push-n73014461986/.
8. Oreskes, Naomi. 'The Scientific Consensus on Climate Change.' *Science*, vol. 306, no. 5702, 2004, p. 1686, JSTOR, www.jstor.org/stable/3839754.
9. Cook, John, et al. 'Quantifying the Consensus on Anthropogenic Global Warming in the Scientific Literature.' Environmental Research Letters, vol. 8, no. 2, 15 May 2013, iopscience.iop.org/article/10.1088/1748-9326/8/2/024024;jsessionid=7E6D46658D8BDE58FDD7F91D9C206BCA.c2.iopscience.cld.iop.org.
10. Friedman, David. 'A Climate Falsehood You Can Check for Yourself.' *Ideas*, Blogger, 16 Feb. 2014, daviddfriedman.blogspot.com/2014/02/a-climate-falsehood-you-can-check-for.html.
11. Epstein, Alex. '"97% Of Climate Scientists Agree" Is 100% Wrong.' *Forbes*, Forbes Magazine, 9 Jan. 2015, www.forbes.com/sites/alexepstein/2015/01/06/97-of-climate-scientists-agree-is-100-wrong/#3e8897023f9f.
12. Hansen, James, et al. 'Whither U.S. Climate?' *Science Briefs*, NASA, Aug. 1999, www.giss.nasa.gov/research/briefs/hansen_07/.
13. 'Annual Mean Temperature Change in the United States.' GISS Surface Temperature Analysis, NASA, 13 Oct. 2016, data.giss.nasa.gov/gistemp/graphs_v3/.
14. Wallace, James P., et al. 'On the Validity of NOAA, NASA and Hadley CRU Global Average Surface Temperature Data & the Validity of EPA's CO2 Endangerment Finding,' 27 June 2017, thsresearch.files.wordpress.com/2017/05/ef-gast-data-research-report-062717.pdf.
15. Ibid.
16. Watts, Anthony. 'Uh, Oh – Raw Data in New Zealand Tells a Different Story than the "Official" One.' *WUWT*, Watts Up With That?, 25 Nov. 2009, wattsupwiththat.com/2009/11/25/uh-oh-raw-data-in-new-zealand-tells-a-different-story-than-the-official-one/.
17. Schilling, Chelsea. 'History of Climate Gets "Erased" Online.' *WND*, WND, 21 Dec. 2009, www.wnd.com/2009/12/119745/.
18. Bowater, Donna. 'Climate Change Lies Are Exposed.' *Sunday Express*, Express.co.uk, 30 Aug. 2010, www.express.co.uk/news/uk/196642/Climate-change-lies-are-exposed.
19. 'Climate Change Indicators: High and Low Temperatures.' *Climate Change Indicators*, Environmental Protection Agency, 17 Dec. 2016, www.epa.gov/climate-indicators/climate-change-indicators-high-and-low-temperatures.
20. Lochbaum, David. 'Nuclear Power in the Future: Risks of a Lifetime.' *It Is Two and a Half Minutes to Midnight*, Bulletin of the Atomic Scientists, 24 Feb. 2016, thebulletin.org/nuclear-power-future-risks-lifetime9185.
21. Ibid.
22. Donn, Jeff. 'Safety Rules Loosened for Aging Nuclear Reactors.' *NBCNews.com*, NBCUniversal News Group, 20 June 2011, www.nbcnews.com/id/43455859/ns/us_news-environment/t/safety-rules-loosened-aging-nuclear-reactors/#.WZy3-Weotkh.
23. Lochbaum, n 20 above.
24. The Associated Press. 'Could Aging N.J. Nuclear Power Plant Be Target for Cybersecurity Attack?' *NJ.com*, NJ Advance Media, 19 July 2017, www.nj.com/ocean/index.ssf/2017/07/questions_surround_nuclear_plants_cybersecurity_am.html.
25. '"Nightmare Scenario": Nuclear Power Plants Vulnerable to Hacking by Terrorists.' *Nuclear Terrorism, Fallout Shelters, Radiation Risks, Dirty Bombs, Hacking*, Homeland Security News Wire, 16 Dec. 2016, www.homelandsecuritynewswire.com/dr20161216-nightmare-scenario-nuclear-power-plants-vulnerable-to-hacking-by-terrorists.
26. The Associated Press, n 24 above.
27. Biello, David. 'Spent Nuclear Fuel: A Trash Heap Deadly for 250,000 Years or a Renewable Energy Source?' *Sustainability*, Scientific American, 28 Jan. 2009, www.scientificamerican.com/article/nuclear-waste-lethal-trash-or-renewable-energy-source/.

28. 'How Old Are The Pyramids?.' *NOVA*. PBS, June n.d. Web. 23 Jan. 2016. http://www.pbs.org/wgbh/nova/pyramid/explore/howold.html.
29. 'An Enduring Problem: Radioactive Waste from Nuclear Energy [Point of View].' *Proceedings of the IEEE, Proc. IEEE*, no. 3, 2017, p.415. EBSCOhost, doi:10.1109/JPROC.2017.2661518.
30. Associated Press. 'Population Density around Nuke Plants Soars.' *CBS News*, CBS Interactive, 27 June 2011, www.cbsnews.com/news/population-density-around-nuke-plants-soars.
31. Holdren, John P. 'Threats to Civil Nuclear-Energy Facilities.' *Science and Technology to Counter Terrorism: Proceedings of an Indo-U.S. Workshop*, National Academies Press: OpenBook, 2007, www.nap.edu/read/11848/chapter/8.
32. 'An Enduring Problem', n 29 above.
33. Behrens, Carl, and Mark Holt. 'Nuclear Power Plants: Vulnerability to Terrorist Attack.' *CRS Report for Congress*, Wisconsin State Legislature, 9 Aug. 2005, legis.wisconsin.gov/lc/committees/study/2006/npowr/files/rs21131.pdf.
34. Editorial Staff. 'Targets for Terrorism: Nuclear Facilities.' *Council on Foreign Relations*, Council on Foreign Relations, 1 Jan. 2006, www.cfr.org/backgrounder/targets-terrorism-nuclear-facilities.
35. National Research Council (U.S.) Committee on the Safety and Security of Commercial Spent Nuclear Fuel Storage. Safety and Security of Commercial Spent Nuclear Fuel Storage, Public Report. The National Academies, 2006, www.apcnean.org.ar/arch/e38669d13b132d7e679705af8817f8f4.pdf.

# Index